4
Modern Mathematics for Schools

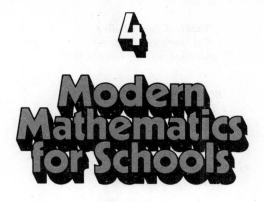

4
Modern Mathematics for Schools

Second Edition
Scottish Mathematics Group

Blackie

Chambers

Blackie & Son Limited
Bishopbriggs · Glasgow
5 Fitzhardinge Street · London W1

W & R Chambers Limited
11 Thistle Street · Edinburgh 2
6 Dean Street · London W1

© *Scottish Mathematics Group 1972*
First Published 1972

All Rights Reserved.
No part of this publication may be reproduced,
stored in a retrieval system, or transmitted,
in any form or by any means,
electronic, mechanical, recording or otherwise,
without prior permission of the Publishers

Designed by James W. Murray

International Standard Book Numbers
Pupils' Book
Blackie 0 216 89412 3
Chambers 0 550 75914 X
Teachers' Book
Blackie 0 216 89413 1
Chambers 0 550 75924 7

Printed in Great Britain by
McFarlane & Erskine Limited · Edinburgh
Set in 10pt Monotype Times Roman

Members associated with this book

W. T. Blackburn
Dundee College of Education

W. Brodie
Trinity Academy

C. Clark
Formerly of Lenzie Academy

D. Donald
Formerly of Robert Gordon's College

R. A. Finlayson
Allan Glen's School

Elizabeth K. Henderson
Westbourne School for Girls

J. L. Hodge
Dundee College of Education

J. Hunter
University of Glasgow

R. McKendrick
Langside College

W. More
Formerly of High School of Dundee

Helen C. Murdoch
Hutchesons' Girls' Grammar School

A. G. Robertson
John Neilson High School

A. G. Sillitto
Formerly of Jordanhill College of Education

A. A. Sturrock
Grove Academy

Rev. J. Taylor
St. Aloysius' College

E. B. C. Thornton
Bishop Otter College

J. A. Walker
Dollar Academy

P. Whyte
Hutchesons' Boys' Grammar School

H. S. Wylie
Govan High School

Only with Chapter 1 of the arithmetic section

R. D. Walton
Dumfries Academy

Contributor of the Computer Studies

A. W. McMeeken
Dundee College of Education

Preface

Book 1 of the original series *Modern Mathematics for Schools* was first published in July 1965. This revised series has been produced in order to take advantage of the experience gained in the classroom with the original textbooks and to reflect the changing mathematical needs in recent years, particularly as a result of the general move towards some form of comprehensive education.

Throughout the whole series, the text and exercises have been cut or augmented wherever this was considered to be necessary, and nearly every chapter has been completely rewritten. In order to cater more adequately for the wider range of pupils now taking certificate-oriented courses, the pace has been slowed down in the earlier books in particular, and parallel sets of A and B exercises have been introduced where appropriate. The A sets are easier than the B sets, and provide straightforward but comprehensive practice; the B sets have been designed for the more able pupils, and may be taken in addition to, or instead of, the A sets. Often from Book 4 onwards a basic exercise, which should be taken by all pupils, is followed by a harder one on the same work in order to give abler pupils an extra challenge, or further practice; in such a case the numbering is, for example, Exercise 2 followed by Exercise 2B. It is hoped that this arrangement, along with the *Graph Workbook for Modern Mathematics*, will allow considerable flexibility of use, so that while all the pupils in a class may be studying the same topic, each pupil may be working examples which are appropriate to his or her aptitude and ability.

Each chapter is backed up by a summary, and by revision exercises; in addition, cumulative summaries and exercises have been introduced at the end of alternate books. A new feature is the series of Computer Topics from Book 4 onwards. These form an elementary introduction to computer studies, and are primarily intended to give pupils some appreciation of the applications and influence of computers in modern society.

Books 1 to 7 provide a suitable course for many modern Ordinary Level and Ordinary Grade syllabuses in mathematics, including the University of London GCE Syllabus C, the Associated Examining Board Syllabus C, the Cambridge Local Syndicate Syllabus C, and the Scottish Certificate of Education. Books 8 and 9 complete the work for the Scottish Higher Grade Syllabus, and provide a good preparation for all Advanced Level and Sixth Year Syllabuses, both new and traditional.

Related to this revised series of textbooks are the *Modern Mathematics Newsletters* (No. 2, February 1972), the *Teacher's Editions* of the textbooks, the *Graph Workbook for Modern Mathematics*, the *Three-Figure Tables for Modern Mathematics*, and the booklets of *Progress Papers for Modern Mathematics*. These new Progress Papers consist of short, quickly marked objective tests closely connected with the textbooks. There is one booklet for each textbook, containing A and B tests on each chapter, so that teachers can readily assess their pupils' attainments, and pupils can be encouraged in their progress through the course.

The separate headings of Algebra, Geometry, Arithmetic, and later Trigonometry and Calculus, have been retained in order to allow teachers to develop the course in the way they consider best. Through-out, however, ideas, material and method are integrated *within* each branch of mathematics and *across* the branches; the opportunity to do this is indeed one of the more obvious reasons for teaching this kind of mathematics in the schools—for it is *mathematics* as a whole that is presented.

Pupils are encouraged to find out facts and discover results for themselves, to observe and study the themes and patterns that pervade mathematics today. As a course based on this series of books progresses, a certain amount of equipment will be helpful, particularly in the development of geometry. The use of calculating machines, slide rules, and computers is advocated where appropriate, but these instruments are not an essential feature of the work.

While fundamental principles are emphasized, and reasonable attention is paid to the matter of structure, the width of the course should be sufficient to provide a useful experience of mathemetics for those pupils who do not pursue the study of the subject beyond school level. An effort has been made throughout to arouse the interest of all pupils and at the same time to keep in mind the needs of the future mathematician.

The introduction of mathematics in the Primary School and recent changes in courses at Colleges and Universities have been taken into account. In addition, the aims, methods, and writing of these books have been influenced by national and international discussions about the purpose and content of courses in mathematics, held under the auspices of the Organization for Economic Co-operation and Development and other organizations.

The authors wish to express their gratitude to the many teachers who have offered suggestions and criticisms concerning the original series of textbooks; they are confident that as a result of these contacts the new series will be more useful than it would otherwise have been.

Algebra

1 Further Sets and Graphs

The description of a set; equal sets; the number of elements in a set; sets contained in sets—subsets; the intersection of sets; the union of sets; the complement of a set; set of real numbers; graphs; summary.

2 Systems of Equations and Inequations in Two Variables 23

The solution set of an equation or inequation in two variables; linear equations and inequations in two variables; the solution set of a system of inequations; the solution set of a system of equations—graphical method, elimination method and substitution method; systems of equations with fractions; problems leading to systems of equations in two variables; summary.

3 Formulae and Literal Equations 40

Constructing and using a formula; revising the methods of solving equations; literal equations; cross-multiplication; changing the subject of a formula; summary.

Revision Exercises 57

Cumulative Revision Section (Books 1-4) 69

Geometry

1 Translation — 89

Position and direction; displacements; translation;
a number pair notation; patterns; two successive translations;
summary.

2 The Calculation of Distance — 110

Calculating the length of a line; Pythagoras' theorem;
using coordinates; the distance formula;
the converse of Pythagoras' theorem; summary.

Topics to Explore — 126

Revision Exercises — 128
Cumulative Revision Section (Books 1-4) — 138

Arithmetic

1 Squares and Square Roots of Numbers — 161

Calculating squares of numbers—using a graph, tables of squares
and a slide rule;
calculating square roots of numbers—using a graph, tables of
square roots, a slide rule and an iterative method; summary.

2   Using a Slide Rule   175

Ruler addition and subtraction; from addition to multiplication; calibrating the multiplication scales; the slide rule; multiplication and division; squares and square roots; rough estimates; successive multiplication and division; summary.

3   The Circumference and Area of a Circle   189

The circumference of a circle; approximations for π; calculations involving circumferences of circles; the area of a circle; calculations involving areas of circles; summary.

4   Statistics–2   203

Measurements, samples and populations; frequency tables; class intervals; histograms; frequency polygons; averages—measures of central tendency; mean, mode, median; mean from a frequency table; summary.

Revision Exercises   222

Cumulative Revision Section (Books 1-4)   231

Computer Studies

1   Flow Charts   255

Constructing a flow chart; decision boxes; loops.

Tables of Squares and Square Roots   264

Answers   273

Notation

Sets of numbers

Different countries and different authors
give different notations and definitions
for the various sets of numbers.
In this series the following are used:

E The universal set

ϕ The empty set

N The set of natural numbers $\{1, 2, 3, ...\}$

W The set of whole numbers $\{0, 1, 2, 3, ...\}$

Z The set of integers $\{..., -2, -1, 0, 1, 2, ...\}$

Q The set of rational numbers

R The set of real numbers

The set of prime numbers $\{2, 3, 5, 7, 11, ...\}$

Algebra

Algebra

Note to the Teacher on Chapter 1

Chapter 1 revises and extends the work done on sets and set operations in Book 1, Chapter 1. New work includes the set operations of union and complementation, and also the informal introduction of the real number system. With the introduction of the real numbers, the scope of graphical representation of sets of numbers and mappings is considerably widened. It will be appreciated that when $x \in R$, a solution set such as $\{x: 2x-1 > 6\}$, or $\{(x, y): 2x-y < 6\}$ is much more meaningful and helpful when shown as a graph.

In the Exercises, familiar principles and ideas are met in new situations and may be seen to be special cases of more general ones. For example, applications of the commutative, associative and distributive laws are recurrent themes. In this way, it will become clear that not only is there a language of sets, but more important, that there is an *algebra of sets*.

The first four sections of the chapter contain ideas which will be familiar to the pupil, but the treatment is more mature and comprehensive than previously. Although some teachers may prefer to cover *Sections* 1 and 2 by class discussion, it should be remembered that there may be pupils who did not quite grasp the fundamentals of the subject in the earlier chapter in Book 1. These pupils now have a second chance. Some of the difficulties that may arise are mentioned below:

(1) Confusion of the symbols $\in$ and $\subset$. The pupil should understand clearly that $\in$ relates an element to its set, and $\subset$ relates *sets*. For example, if $A = \{2, 3, 5, 7, 11\}$, $2 \in A$ and $\{2\} \subset A$.
(2) Confusion of part of an element with the whole. For example, if $T = \{\text{isosceles triangles}\}$, the sides of the triangles are *not* members of T.
(3) Repetition of elements in listing. For example, if $S = \{\text{numerals in the number } 346443\}$, then $S = \{3, 4, 6\}$ since there is only one numeral 3 and only one numeral 4 in the arabic system.
(4) The order of the elements in the listing. For example, $\{3, 4, 5\} = \{5, 3, 4\}$ since we are concerned only with whether or not a given object is in the set.

In revising the idea of subsets of a set, it is a good idea to choose a universal set whose elements have plenty of properties. The class being taught provides a suitable set for this purpose! Instances of subsets will then come fast and furious.

Sections 5 and 6 deal with the new topics of union and complement. Class exercises on these topics are carefully graded from the very easy to those requiring more sustained thought. Considerable use is made of the Venn diagram to illuminate the relations between sets. Pupils should be encouraged to make use of coloured pencils to shade the regions showing the relevant sets or to outline the closed curves.

In tackling problems involving the number of elements in subsets, it is recommended that sample questions be worked out on the blackboard as a pupil-teacher activity, so that the pupil will appreciate the method of solving them. With the better classes the derivation of the formula:

$$n(A \cup B \cup C) = n(A)+n(B)+n(C)-n(A \cap B)-n(A \cap C) \\ -n(B \cap C)+n(A \cap B \cap C),$$

will prove to be a stimulating exercise. This result may be illustrated by a Venn diagram or by deduction from the formula for $n(A \cap B)$. An abundance of teaching material may be found in *Sets, Relations and Functions* by Selby and Sweet (McGraw-Hill).

Section 7 introduces the real number system and the real line. At this stage in mathematics, a formal approach is out of the question, so the introduction is made on the grounds of plausibility and the growing mathematical experience of the pupil. The first appeal comes from geometry through applications of Pythagoras' theorem, and the second from the transcendental number π which the pupil meets in arithmetic. The usefulness of the real number system and the real line for one and two-dimensional graphs, intervals, graphs of solution sets of inequations, and mappings, becomes patently clear from the exercises given. To assist the pupil in drawing the graph of a mapping from R to R, a subset of the domain of the mapping is given, the elements of this subset being rational numbers. A suggested scale is also given for each mapping.

Further Sets and Graphs

1 The description of a set

A *set* is a well-defined collection of objects, called its *members* or *elements*. The term 'well-defined' means that it is possible to tell whether a given object belongs to the set or not.

A set may be defined by describing it, by listing its members, or by using set-builder notation.

Examples

1. The set of even whole numbers
 $= \{0, 2, 4, 6, \ldots\}$
 $= \{x : x \text{ is an even number}, x \in W\}$

2. V is the set of vowels in the English alphabet.
 $V = \{a, e, i, o, u\}$

3. $K = \{x : -3 < x < 5, x \in Z\}$
 $= \{-2, -1, 0, 1, 2, 3, 4\}$

4. $L = \{(x, y) : x + y = 6, x, y \in N\}$,
 i.e. the set of points (x, y) such that $x + y = 6$, where x and y are variables on the set of natural numbers.
 $L = \{(1, 5), (5, 1), (2, 4), (4, 2), (3, 3)\}$

Membership of a set is indicated by the symbol $\in$ and *non-membership* by the symbol $\notin$.

$$a \in \{a, e, i, o, u\} \qquad b \notin \{a, e, i, o, u\}$$

Exercise 1 (mainly revision)

1. Is each of the following a well-defined set? If so, describe it, or list its members, or explain how the list could be obtained.
 a. The set of members of your class who are over 150 cm tall.
 b. The set of all factors of 12.
 c. A flock of starlings.

Algebra

   d The set of names of people in your town or county who had a vote in the last general election.
   e The set of fair-haired pupils in your school.
   f The set of even numbers between 1 and 29 which are exactly divisible by 3.

2 Give two examples of well-defined sets by describing the members.

3 Give two examples of well-defined sets by listing the members.

4 Give two examples of well-defined sets, using set-builder notation.

5 Give two examples of collections of objects which do not form sets because they are not well defined.

6 A is the set of whole numbers from 0 to 10 inclusive.
B is the set of prime numbers between 0 and 10.
C is the set of the first six even natural numbers.
D is the set of multiples of 3 between 2 and 10.

   a List the members of each set.
   b Use the symbols $\in$ and $\notin$ to show whether or not the numbers 2, 3, 5, 10 are members of each set.

7 $A = \{1, 2, 3, 4, 5, 6\}$, $B = \{2, 4, 6, 8, 10\}$, $C = \{3, 5, 7, 9, 11\}$.
List the sets given by:

   a $\{x: x > 3, x \in A\}$ *b* $\{x: x \leq 6, x \in B\}$
   c $\{x: 5 < x < 12, x \in C\}$ *d* $\{x: x = 2y+1, y \in B\}$
   e $\{(x, y): x = y, x, y \in B\}$ *f* $\{(x, y): x = 2y, x \in B, y \in C\}$

8 State whether each of the following is true or false for the sets in question 7.

   a $2 \in A$ *b* $2 \in C$ *c* $2 \notin C$ *d* $5 \in B$ *e* $9 \in C$

9 List the sets defined by:

   a $S = \{x: x > 2, x \in N\}$ *b* $T = \{x: x \geq 6, x \in N\}$
   c $A = \{x: -3 < x < 3, x \in Z\}$
   d $B = \{x: x \text{ is a month with less than 31 days}\}$

10 Express in set-builder notation the set of natural numbers

   a greater than 5   *b* less than 10   *c* between 3 and 11
   d between 0 and 5   *e* which are even   *f* which are prime.

2 Equal sets; the number of elements in a set

Two sets A and B are equal when they have exactly the same elements. This means that every member of A is a member of B, and every member of B is a member of A. We then write $A = B$.

Examples

1. $\{\Delta, *, 0\} = \{0, \Delta, *\}$ 2. $\{1, 3, 4, 2\} = \{1, 2, 4, 3\}$

The number of distinct elements in a set A is written $n(A)$.
For example, if $A = \{0, 3, 6, 9, 12\}$, then $n(A) = 5$.
For the empty set ø, $n(ø) = 0$.
Note that if $S = \{1, 2, 3, 4\}$ and $T = \{p, q, r, s\}$, $n(S) = n(T)$ but $S \neq T$. Why not?
A set such as W, the set of all the whole numbers, which contains an *unlimited* collection of elements, is called an *infinite set*. We write:

$$W = \{0, 1, 2, ...\},$$

the three dots indicating 'and so on'.

Exercise 2

1 a List the members of the following sets:
   $F = \{\text{odd whole numbers from 1 to 9 inclusive}\}$
   $G = \{\text{prime numbers between 10 and 24}\}$
   $H = \{\text{multiples of 7 between 1 and 50}\}$
  b Give the values of $n(F)$, $n(G)$ and $n(H)$. Which sets have the same number of elements? Are these sets equal?

2 If $A = \{1, 2, 3, 4\}$ and $B = \{1, 2, 3, 4, 5\}$, then every member of A is a member of B. Why are A and B not equal sets? Give $n(A)$ and $n(B)$.

3 If C is the set of integers between 9528 and 9540 which have 3 in the tens place, find $n(C)$.

4 a If $n(A) = n(B)$, must $A = B$?
  b If $A = B$, must $n(A) = n(B)$?

Algebra

5 $A = \{0, 1, 3, 5\}$, $B = \{5, 3, 1\}$, $C = \{0\}$, $D = \{1\}$, $P = \{\ \}$, and $Q = \{3, 1, 5\}$.
  * a Write down the number of elements in each set.
  * b Write down a pair of equal sets from the given sets.

6 5, 4 and 1 are used to make 3-digit numbers, each being used only once in each number. List the set S of 3-digit numbers, and write down $n(S)$.

7 R, S, I, T, P, H are the sets of ways in which a given rectangle, square, isosceles triangle, equilateral triangle, parallelogram and regular hexagon respectively can fit into their outlines.
  Write down $n(R)$, $n(S)$, $n(I)$, $n(T)$, $n(P)$, $n(H)$.

8 P is the set of prime factors of 2310, and Q is the set of prime numbers less than 13. Show that $P = Q$.

9 Classify each of the following sets as finite or infinite:
  * a The set of telephone subscribers listed in the telephone directory for your area.
  * b $\{2, 4, 6, 8, \ldots, 1\,000\,000\}$ c $\{2, 4, 6, 8, \ldots\}$
  * d The set of even prime numbers which are greater than 3.
  * e $\{n: n > 5, n \in Z\}$ f The set of all points on a line 10 cm long.

10 Write in concise form each of the following sets, by listing the necessary members:
  * a The set of all positive integers less than 100 which are divisible by 3.
  * b The set of all positive odd integers.
  * c The set of all negative integers greater than -25.
  * d The set of reciprocals of the positive integers.

3 Sets contained in sets—subsets

Figure 1 shows a set X of geometrical figures. Using the *properties* of the elements of X we can define sets A, B and C such that:
  * *(1)* A is the set of figures having only *one* axis of symmetry.
  * *(2)* B is the set of figures having exactly *two* axes of symmetry.
  * *(3)* C is the set of figures having *two or more* axes of symmetry.

Sets contained in sets—subsets

$$X = \{\square, \Diamond, \frown, \triangle, \triangle\!\!\!\!\triangle, \hexagon, \times\!\!\!\!\times,)\}$$
  P Q R S T U V W

Listing the elements of *A*, *B* and *C*,

$$A = \{R, T, W\}; \quad B = \{P, Q, V\}; \quad C = \{P, Q, S, U, V\}.$$

Since *X contains A*, *B* and *C* then *A*, *B* and *C* are called *subsets* of *X*, and we write $A \subset X$, $B \subset X$, $C \subset X$. Note that *C* contains *B*, and so $B \subset C$. Try to illustrate these relations by a Venn diagram.

Any suitable set which contains all the sets we are talking about can be taken as the *universal set*, often denoted by *E* '(Entirety'). The sets under discussion are all subsets of *E*.

A set *B* is a *subset* of a set *A* if every member of *B* is a member of *A*. We write $B \subset A$.

This subset relation is illustrated in Figure 2.

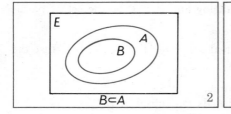

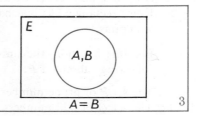

Example. The subsets of $\{a, b\}$ are $\{a\}$, $\{b\}$, $\{a, b\}$ and ø.
Notice that the set itself and the empty set are taken to be subsets of the given set.

If *A* and *B* are two sets such that $A \subset B$ and $B \subset A$, then it follows from the definition that $A = B$. See the Venn diagram in Figure 3.

Exercise 3

1. Put the symbol $\subset$, or the symbol $=$, between as many pairs of the following as you can:

$A = \{0, 1, 2\}$, $B = \{0\}$, $C = \{1, 2\}$, $D = \{2, 1\}$, $F = \{\ \}$, $G = \{1, 2, 3\}$.

Algebra

2 State whether each of the following is true or false:

- a $\{2\} \subset \{0, 2\}$
- b $\{2, 0\} \subset \{2\}$
- c $\emptyset \subset \{1\}$
- d $\{1\} \subset \{1\}$
- e $2 \in \{1, 2, 3\}$
- f $2 \subset \{1, 2, 3\}$

3 The set of months (M) in a year may be written:

$M = \{$Jan, Feb, Mar, Apr, May, Jun, Jul, Aug, Sep, Oct, Nov, Dec$\}$
List the following subsets of M:

- a The set of months having more than 30 days.
- b The set of months having 30 days.
- c The set of months having less than 30 days.

4 Suggest possible universal sets for each of the following:

- a The set of pupils in your class.
- b $\{-3, -1, 0, 1, 2, 3, 4\}$
- c $\{-\frac{7}{8}, -\frac{3}{4}, -\frac{1}{2}, -\frac{1}{4}, 0\}$
- d $\{$kg, g, mg$\}$

5 Use the subset symbol to connect the sets in the following statements:

- a The set of squares (S) is a subset of the set of rectangles (R).
- b All prime numbers (P) are integers (Z).
- c Every rectangle (R) is a parallelogram. (Let P be the set of parallelograms.)

6 For this question, the universal set $E = \{1, 2, 3, \ldots, 10\}$. List the following subsets of E:

- a The set P of prime numbers in E.
- b The set Q of odd numbers in E. Is P a subset of Q?
- c The set S such that $x + 1 > x$, $x \in E$.
- d $\{x: x + 4 < 5, x \in E\}$.

7 a List all the subsets of (*1*) $\{a\}$ (*2*) $\{a, b\}$ (*3*) $\{a, b, c\}$ (*4*) $\{a, b, c, d\}$

- b How many subsets has (*1*) $\{a, b, c, d, e\}$ (*2*) $\{a, b, c, d, e, f\}$?
- c How many subsets does a set with n members have?
- d A certain set has 128 subsets. How many elements are in the set?

8 a If $P \subset Q$ and $Q \subset P$, what can you say about the sets P and Q?

- b What is the name of the set that has only one subset?
- c A, B, C are sets such that $A \subset B$ and $B \subset C$. What can you say about A and C? Illustrate your answer by a Venn diagram.
- d If $A \subset B$ and $C \subset B$, does it follow that $A \subset C$? Illustrate by a Venn diagram.

4 The intersection of sets

The *intersection* of two sets A and B is the set of elements which are members of A and are also members of B.
This new set is written A ∩ B ('A intersection B').
Example 1. If $A = \{p, q, r, s\}$ and $B = \{r, s, t\}$, then $A \cap B = \{r, s\}$. The elements in the shaded region in Figure 4 represent $A \cap B$.

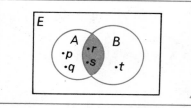
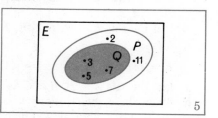

Example 2. If the universal set E is the set of whole numbers, P = {prime numbers less than 12}, Q = {odd numbers between 2 and 8}, list the sets P, Q and P ∩ Q. Illustrate by a Venn diagram.

$P = \{2, 3, 5, 7, 11\}$, $Q = \{3, 5, 7\}$, $P \cap Q = \{3, 5, 7\}$.

The intersection of P and Q is shown by the elements in the shaded region in the Venn diagram in Figure 5.
Note that $Q \subset P$, which is shown in the diagram by the fact that set Q is entirely contained in set P.

Exercise 4

1 $E = \{1, 2, 3, \ldots, 10\}$; $A = \{1, 2, 3, 4, 5\}$ and $B = \{4, 5, 6, 7\}$.
   a Write out in list form the sets $A \cap B$ and $B \cap A$.
   b What do you notice about these two sets? Which law for intersection does this result suggest?
   c Draw a Venn diagram to illustrate the intersection of the sets.

2 E is the set of positive integers; $P = \{1, 2, 3\}$ and $Q = \{1, 2, 3, 4, 5, 6, 7\}$.
   a List $P \cap Q$ and $Q \cap P$.
   b What letter denotes a set equal to $P \cap Q$?
   c Illustrate the relationships between the sets by a Venn diagram.

Algebra

3   $E = \{1, 2, 3, \ldots\}$; $C = \{1, 3, 5, 7\}$ and $D = \{2, 4, 6, 8\}$.
  a Find $C \cap D$. What set is equal to $C \cap D$?
  b Illustrate by a Venn diagram.

   Note: Two sets which have no members in common are said to be *disjoint*. C and D are disjoint.

4   A is the set of the first five positive odd numbers; $B = \{7, 9, 11\}$, $C = \{1, 2, 3, 4\}$. List the sets A, $A \cap B$, $A \cap C$ and $B \cap C$.

5   $E = \{$letters of the alphabet$\}$; $P = \{a, b, c, d, e, f\}$, $Q = \{b, c, d\}$, $R = \{d, e, f\}$, $S = \{f\}$.
   List the sets $P \cap Q$, $Q \cap R$, $R \cap S$, $P \cap R$, $Q \cap S$, $P \cap P$, $Q \cap Q$, $S \cap S$.

6   $E = \{1, 2, 3, 4, 5, 6\}$; $A = \{1, 2, 3, 4\}$, $B = \{3, 4, 5\}$, $C = \{5\}$.
  a List the sets $A \cap B$, $B \cap C$, $A \cap C$, $E \cap B$.
  b Show these intersections in separate Venn diagrams. The use of shading or colouring will help to show the intersections.

7   $E = \{$all triangles$\}$, $I = \{$isosceles triangles$\}$, $K = \{$right-angled triangles$\}$. Represent these sets in a Venn diagram and explain the reason for showing $I \cap K$ in the way that you did.

8   X is the set of multiples of 6, less than 35. Y is the set of multiples of 8, less than 35. List the members of X, Y and $X \cap Y$. Ignoring zero in $X \cap Y$ we obtain the LCM of 6 and 8. Write down this LCM.

9   Repeat question *8* to find the LCM of:
  a 4 and 6. (Take multiples up to 30.)
  b 9 and 12. (Take multiples up to 50.)

10 If S is the set of roses in a flower shop and C is the set of yellow flowers in the shop, describe in words the set $S \cap C$. In what circumstances is $S \cap C = \emptyset$?

Exercise 4B

1   $E = \{1, 2, 3, \ldots, 10\}$; $A = \{1, 2, 3, 4, 5\}$, $B = \{2, 4, 5, 6\}$, $C = \{2, 4, 6, 8\}$.
  a List the sets: (*1*) $A \cap B$ (2) $B \cap C$ (3) $(A \cap B) \cap C$ (4) $A \cap (B \cap C)$.
  b What can you deduce from the answers to (*3*) and (*4*)?

The intersection of sets

2 The Venn diagram in Figure 6 shows the intersections of three sets A, B and C. Make four copies of this diagram.

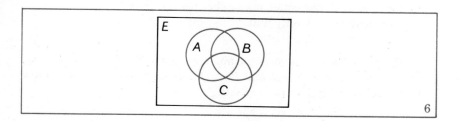

Figure 6

a In the first copy, shade the region showing $A \cap B$; in the second copy, shade the region showing $(A \cap B) \cap C$. Name each diagram by the set shown.

b Repeat for the sets $B \cap C$ and $A \cap (B \cap C)$, using the remaining two copies.

c Does it make sense to write $A \cap B \cap C$? We have seen that the commutative law holds for the intersection of sets. From the results obtained in questions *1* and *2*, what other law appears to hold for intersection?

3 In a Fourth Form of 20 boys, 15 take mathematics, 12 take science, and 10 take mathematics and science. Show this information in a Venn diagram. How many boys take neither mathematics nor science?

4 All the pupils in a class of 35 take French or German, or both. 29 take French and 15 take German. Show this information in a Venn diagram, marking x for the number of pupils who take both languages. Form an equation in x, and find x.

5 In a survey of 120 pupils it was found that 17 had no liking for music. Of the rest, 88 said they liked 'pop' music and 26 said they liked classical music. Illustrate this information in a Venn diagram, marking x for the number who liked both 'pop' and classical music. Form an equation in x, and find x.

6 A survey of 180 second-year pupils showed that 100 drank tea, 92 drank coffee, 115 drank milk, 43 drank tea as well as milk, 52 drank coffee as well as tea, 57 drank coffee as well as milk, and 25 drank all three.

Algebra

a Using *T*, *C* and *M* to represent the sets of tea, coffee and milk drinkers, show this information in a Venn diagram by first marking in the number who drank all three (see Figure 6).

b How many drank coffee but neither tea nor milk?

c How many drank only milk?

d How many drank only tea?

7 *P*, *Q*, and *R* are three non-empty sets with the properties $P \subset Q$, $Q \cap R \neq \emptyset$, $P \cap R = \emptyset$ and $R \not\subset Q$. Draw a Venn diagram to show the relationship between the sets.

5 The union of sets

The *union* of two sets *A* and *B* is the set of elements which belong either to *A* or to *B*, or to both *A* and *B*.

This new set is written $A \cup B$ ('*A* union *B*', or 'the union of *A* and *B*').

Example 1. If $E = \{1, 2, 3, \ldots, 10\}$, $A = \{1, 2, 3, 4, 5\}$ and $B = \{4, 5, 6, 7\}$, then $A \cup B = \{1, 2, 3, 4, 5, 6, 7\}$.

The shaded region in the Venn diagram in Figure 7 represents $A \cup B$. Notice that $A \cap B = \{4, 5\}$.

Example 2. If $P = \{2, 4, 6, \ldots\}$ and $Q = \{1, 3, 5, \ldots\}$, then $P \cup Q = \{1, 2, 3, \ldots\}$, the set of natural numbers. Notice that $P \cap Q = \emptyset$.

The shaded region in the Venn diagram in Figure 8 represents $P \cup Q$ for some suitable universal set *E*. (Suggest an appropriate universal set *E*.)

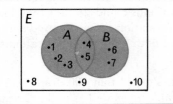

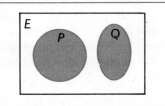

The union of sets

Exercise 5

1. **a** Using Figure 9, write in list form $A \cup B$ for each of the cases shown.
   b Is $A \cup B = B \cup A$ in each case?
   c List also the members of $A \cap B$ for each case.

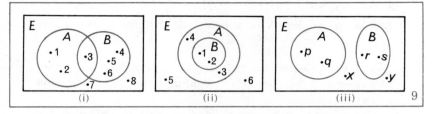

2. **a** Copy the Venn diagrams shown in Figure 10.
   b Using vertical lines, shade in each case the region which shows $P \cup Q$.
   c Using horizontal lines, shade in each case the region which shows $P \cap Q$.
   d Is $P \cup Q = Q \cup P$? Which law for union does this suggest?

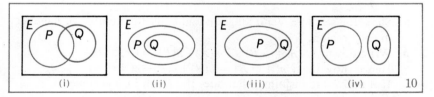

3. List the members of the *union* of each of the following pairs of sets:
   a $P = \{1, 2, 3\}$, $Q = \{2, 3, 5, 7\}$ **b** $X = \{0, 1, 2\}$, $Y = \{3, 4, 5\}$
   c $S = \{a, b, c\}$, $T = \{a, b\}$ **d** $A = \{a, b, c\}$, $B = \{a, d, e\}$
   Illustrate each of your answers by a Venn diagram.

4. $A = \{a, b, c\}$, $B = \{c, d, e\}$. Show the sets A and B in a Venn diagram, and shade the region representing $A \cup B$. What is $n(A \cup B)$?

5. If $P = \{$odd numbers from 1 to 7 inclusive$\}$, and $Q = \{$even numbers from 2 to 8 inclusive$\}$, list $P \cup Q$. Illustrate in a Venn diagram. What is $n(P \cup Q)$?

6. If S is the set of pupils in your class and T is the set of pupils in the other classes in your school, describe $S \cup T$.

Algebra

7 $V = \{\text{vowels in the alphabet}\}$, $C = \{\text{consonants}\}$.
  a Describe the sets *(1)* $V \cup C$ *(2)* $V \cap C$.
  b Find *(1)* $n(V \cup C)$ *(2)* $n(V \cap C)$.

8 $X = \{\text{prime factors of 60}\}$, $Y = \{\text{the first four prime numbers}\}$. Is $X = Y$? List the sets $X \cup Y$ and $X \cap Y$.

9 If $R = \{\text{rectangles}\}$, $S = \{\text{squares}\}$, describe the following sets:
  a $R \cup S$ **b** $R \cap S$

10 Draw a Venn diagram for each of the following, and enter the number of elements in the appropriate regions. Then calculate the number of elements asked.
  a $n(A) = 50$, $n(B) = 62$, and $n(A \cap B) = 26$. Calculate $n(A \cup B)$.
  b $n(X) = 7$, $n(Y) = 11$, and X and Y are disjoint. Calculate $n(X \cup Y)$.
  c $n(P) = 23$, $n(Q) = 25$, and $P \subset Q$. Calculate $n(P \cup Q)$.

Exercise 5B

1 A and B are sets such that $n(A) = p+q$, $n(B) = q+r$ and $n(A \cap B) = q$.
  a Show the sets in a Venn diagram and enter the number of members in each region.
  b Write down: *(1)* $n(A \cup B)$ *(2)* $n(A)+n(B)-n(A \cap B)$
    Hence show that: $n(A \cup B) = n(A)+n(B)-n(A \cap B)$
  c If A and B are disjoint, what does the result in **b** become?

2 Each of the following diagrams shows the relationships between sets A, B and C. Describe the set indicated by the shaded region in terms of unions and/or intersections of A, B and C.

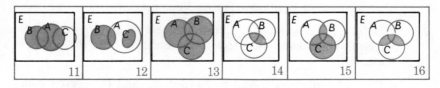

    11   12   13   14   15   16

3 $A = \{0, 1, 2, 3\}$, $B = \{1, 3, 5, 7\}$, $C = \{2, 3, 5, 8\}$.
  List each of the following sets:
  a $A \cup B$ **b** $A \cup C$ **c** $B \cup C$ **d** $A \cup A$ **e** $A \cup \emptyset$
  f $A \cap (A \cup B)$

The union of sets

4 Using the sets in question *3*, list the following sets:
   a $(A \cup B) \cup C$ *b* $A \cup (B \cup C)$
What do you notice about your answers?

5 Make four copies of the diagram shown in Figure 6 on page 11.
   a In the first copy, shade the region showing $A \cup B$; in the second copy, shade the region showing $(A \cup B) \cup C$. Name each diagram.
   b Repeat for the sets $B \cup C$ and $A \cup (B \cup C)$.
   c Does it make sense to write $A \cup B \cup C$? What law appears to hold for the union of three sets? Check with your answers in question *4*.

6 $A = \{1, 2, 3, 4, 5, 6\}$, $B = \{2, 4, 6, 8, 10\}$, $C = \{3, 6, 9, 12\}$.
   a Write each of the following sets in list form:
     (*1*) $A \cup B$ (*2*) $B \cup C$ (*3*) $A \cup C$ (*4*) $A \cap B$ (*5*) $B \cap C$
     (*6*) $A \cap C$
   b Use these sets to obtain in list form:
     (*1*) $A \cap (B \cup C)$ (*2*) $A \cup (B \cap C)$ (*3*) $(A \cap B) \cup (A \cap C)$
     (*4*) $(A \cup B) \cap (A \cup C)$
   c In *b* which sets are equal?

7 Draw five Venn diagrams similar to that in Figure 6, page 11.
   a In the first diagram, shade the region showing $B \cup C$; in the second, show $A \cap (B \cup C)$; in the remaining three, show in succession the sets $A \cap B$, $A \cap C$, and $(A \cap B) \cup (A \cap C)$. Which sets are equal?
   b Repeat the above for the sets $B \cap C$, $A \cup (B \cap C)$, $A \cup B$, $A \cup C$, and $(A \cup B) \cap (A \cup C)$. Which of these sets are equal?

Note: Questions *6* and *7* illustrate the *two distributive laws* for sets:

(1) $A \cap (B \cup C) = (A \cap B) \cup (A \cap C)$
(2) $A \cup (B \cap C) = (A \cup B) \cap (A \cup C)$

i.e. intersection is distributive over union, and union is distributive over intersection.

8 Draw a Venn diagram to show the intersections of three non-empty sets A, B, and C. Use the diagram to help you solve the following:

Given that $n(A) = 20$, $n(B) = 18$, $n(C) = 25$, $n(A \cap B) = 8$, $n(A \cap C) = 9$, $n(B \cap C) = 11$, and $n(A \cap B \cap C) = 5$, find $n(A \cup B \cup C)$.

6 The complement of a set

Consider the universal set $E = \{1, 2, 3, 4, 5\}$. $A = \{2, 4, 5\}$ is a subset of E. All the members of E that are *not* in A make the subset $\{1, 3\}$. The subset $\{1, 3\}$ is called the *complement* of A with respect to E, and is denoted by A'. See the Venn diagram in Figure 17.

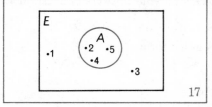
17

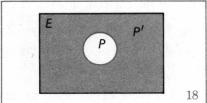
18

Example. If $E = \{\text{whole numbers}\}$, $P = \{\text{even whole numbers}\}$, then P', i.e. the complement of P, $= \{\text{odd whole numbers}\}$. See the Venn diagram in Figure 18.

If A is a subset of a universal set E, then the *complement of A* with respect to E is the set of elements in E which do not belong to A. The complement of A with respect to E is denoted by A'.
It follows that:

(i) $A \cup A' = E$ (ii) $A \cap A' = \emptyset$ (iii) $(A')' = A$.

Exercise 6

1 $E = \{p, q, r, s, t, v\}$ and $X = \{p, r, v\}$. List the members of X'.

2 $E = \{1, 2, 3, \ldots, 20\}$. P is the set of prime numbers in E.
List the members of P and P'.

3 $W = \{\text{whole numbers}\}$ and subset $T = \{\text{even numbers}\}$. Describe in words the complement of T with respect to W.

4 The universal set is the set of all integers. What is the complement of the set of negative integers?

The complement of a set

5. E is the set of all triangles. A is the set of all isosceles triangles. Describe the set A'.

6. $E = \{0, 1, 2, \ldots, 10\}$. If $A = \{x: x \geqslant 2, x \in E\}$ and $B = \{x: x < 5, x \in E\}$, give A' and B' in set-builder form.

7. $E = \{$rectangles, squares, parallelograms, rhombuses, kites$\}$.
   $A = \{$quadrilaterals with bilateral symmetry$\}$, $B = \{$quadrilaterals with half turn symmetry$\}$, and A and B are subsets of E.
   List A, A', B and B'.

8. $E = \{1, 2, 3, \ldots, 8\}$, $A = \{1, 2, 3, 5, 7\}$, $B = \{1, 3, 5, 7\}$, $C = \{2, 4, 6, 8\}$. List the sets A', B' and C'.
   State whether each of the following is true or false:

   a $A \cap A' = \emptyset$ b $A \cup A' = \emptyset$ c $A \subset B$ d $B \subset A$
   e $A' \subset B'$ f $B' \subset A'$ g $B \subset C$ h $B \subset C'$

9. Using the information in Figure 19, list the following sets:

   a A b A' c B d B' e $A \cap B$
   f $(A \cap B)'$ g $A \cup B$ h $(A \cup B)'$ i $A' \cup B'$ j $A' \cap B'$
   Which two pairs of the above sets are equal?

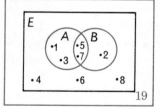

19

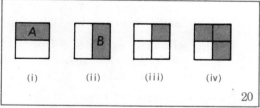

(i)　　(ii)　　(iii)　　(iv)

20

10. In Figure 20 the universal set is the set of all points in the large square. Subsets A and B are the points in the shaded regions shown in (i) and (ii). In terms of the complements, intersections and unions of A and B, name the unshaded regions in (i) and (ii), and the shaded regions in (iii) and (iv).

Algebra

Exercise 6B

1. From the information shown in Figure 21 list the elements of:

   a A b A' c B d B' e $A \cap B'$ f $B \cap A'$

   Copy Figure 21, and shade the regions representing $A \cap B'$ and $B \cap A'$. List the elements of the set $(A \cap B') \cup (B \cap A')$.

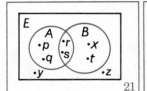
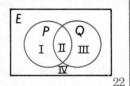
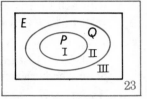

2. Copy and complete the table for the sets illustrated in Figure 22:

Region	I	II	III	IV
Set description	$P \cap Q'$	...	...	...

3. Repeat question 2 for the sets illustrated in Figure 23.

4. In question 2, $P = \{$quadrilaterals which have bilateral symmetry$\}$,
   $Q = \{$quadrilaterals which have half turn symmetry$\}$.
   Describe *in words* each of the sets represented by the regions I, II, III, IV, and give an example of a member of each set.

5. Using set-builder notation, $A \cup B = \{x: x \in A \text{ or } x \in B\}$, where *or* is used in its *inclusive* sense, i.e. $x \in A$, or $x \in B$, or $x \in A$ *and* B. In a similar way, define: a A' b $A \cap B$ c $A \cap B'$

6. Copy diagram (i) of Figure 10 on page 13 for each of the following, and by shading show on it: a $(P \cup Q)'$ b P' c Q' d $P' \cap Q'$
   Which two sets are equal?

7. Repeat question 6 for the sets:

   a $(P \cap Q)'$ b P' c Q' d $P' \cup Q'$

   Note: Questions 6 and 7 verify De Morgan's laws which are as follows:

   (1) $(A \cup B)' = A' \cap B'$ and   (2) $(A \cap B)' = A' \cup B'$

   In words, the complement of the union of two sets A and B equals the intersection of their complements. In the same way, can you express the second law in words?

7 The set of real numbers (R); graphs

So far, we have met the sets of natural numbers (*N*), whole numbers (*W*), integers (*Z*) and rational numbers (*Q*), and we have represented numbers from each set by points on a number line. Each of these numbers corresponds to a definite point on the number line. The question now is: 'Does each point on the number line correspond to a number in one of the above sets?' Look at the following.

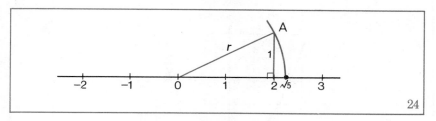

24

From Figure 24, using Pythagoras' theorem,
$$r^2 = 2^2 + 1^2 = 5$$
Hence $r = \sqrt{5} = 2\cdot236\ldots$, a number which cannot be expressed as a ratio of two integers or as a decimal. Therefore $\sqrt{5}$ does not belong to *Q*.

$\sqrt{5}$ is called an *irrational* number. We can find its approximate position on the number line by drawing an arc of a circle with centre O and radius OA as shown. The number π is another example of a number which does not belong to *Q*. It lies between 3·141 and 3·142, and an approximation to its value has been worked out to over 100 000 decimal places by computer.

The set of numbers which is in one-to-one correspondence with the set of all points on the number line is called *the set of real numbers* and is denoted by *R*. Real numbers which are not rational are called irrational numbers.

The line on which we mark or graph the set of real numbers is called the real number line, or simply 'the real line'. It is the *solid* line of coordinate geometry.

The set of real numbers, together with the operations of addition and multiplication, is called the *real number system*.

Algebra

In Book 3 Algebra, Chapter 1, we drew graphs of mappings on the set of integers. We shall now show how to draw the graphs of mappings from R to R.

Consider the mapping $x \to x^2$ on the set of integers Z. We can draw part of the graph using the ordered pairs of the mapping in the table:

x	-3	-2	-1	0	1	2	3
x^2	9	4	1	0	1	4	9

The corresponding graph is shown by the *dots* in Figure 25.

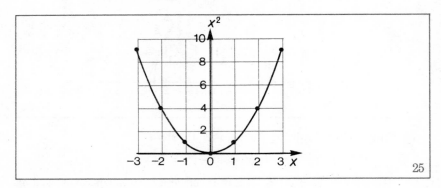

If x is a variable on the set of real numbers R, we have an ordered pair (x, x^2) for each element of the set R, so there are infinitely many of these ordered pairs in the mapping. We cannot list them all in a table, and the most convenient way to show them is by joining up the dots by means of a smooth curve, part of which is shown in Figure 25. The complete smooth curve is the graph of the mapping from R to R.

The set of real numbers (R); graphs

Exercise 7

Draw the graphs of the following mappings from R to R. In each case start by making a table, for which a replacement set for x is given. Then draw the graph on 2-mm squared paper.

1 a $x \to x$ b $x \to -x$ c $x \to 2x$ d $x \to x-2$
   For the table, $x \in \{-3, -2, -1, 0, 1, 2, 3\}$.
   Scales: 1 cm to 1 unit on each axis.

2 $x \to x^2$.   For the table, $x \in \{-4, -3, -2, -1, 0, 1, 2, 3, 4\}$.
   Scales: 2 cm to 1 unit for x, 1 cm to 1 unit for x^2.

3 Repeat question 2 for: a $x \to -x^2$,   b $x \to \tfrac{1}{2}x^2$.

4 $x \to \sqrt{x}$.   For the table, $x \in \{0, 1, 4, 9, 16, 25\}$.
   Scales: 2 cm to 5 units for x, 2 cm to 1 unit for $\sqrt{x}$.

5 $x \to x^3$.   For the table, $x \in \{-3, -2, -1, 0, 1, 2, 3\}$.
   Scales: 2 cm to 1 unit for x, 1 cm to 5 units for x^3.

6 $x \to \dfrac{12}{x}$.   For the table, $x \in \{1, 2, 3, 4, 6, 8, 12\}$.
   Scales: 1 cm to 1 unit on each axis.

7 $x \to -\dfrac{12}{x}$.   For the table, $x \in \{1, 2, 3, 4, 6, 8, 12\}$.
   Scales: 1 cm to 1 unit on each axis.

8 $x \to 4x-x^2$.   For the table, $x \in \{0, 1, 2, 3, 4, 5\}$.
   Scales: 2 cm to 1 unit for x, 1 cm to 1 unit for $4x-x^2$.

Algebra

Summary

1. A *set* is a well-defined collection of objects. $A = \{a, b, c\}$

2. Each object in a set is a *member* or *element* of that set, and *belongs* to the set. $a \in A,\ p \notin A$

3. *Equal sets* have exactly the same members. $B = \{c, a, b\},\ A = B$

4. The *empty set* is the set with no members. $\emptyset$ or $\{\ \}$

5. A *universal set* contains *all* the elements being discussed. E ('Entirety')

6. A set B is a *subset* of a set A if every member of B is a member of A. $B \subset A,\ A \subset A$
 $\emptyset \subset A,\ \emptyset \subset E$

7. The *intersection* of two sets A and B is the set of elements which are members of A and are also members of B. $A \cap B$

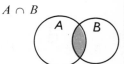

8. The *union* of two sets A and B is the set of elements which belong either to A or to B, or to both A and B. $A \cup B$

9. The *complement* of A with respect to E is the set of elements in E which do not belong to A. $A',\ (A')' = A$

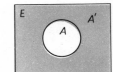

10. The *operations intersection* and *union* of sets are commutative and associative, and each is distributive over the other.

Note to the Teacher on Chapter 2

Chapter 2 draws together several important ideas and techniques: the use of the coordinate plane; the handling of equations and inequations; calculations arising from positive and negative replacements for x and y; the formation of equivalent equations by adding the same number to each side or multiplying each side by the same number; and setting up a simple mathematical model for the solution of a problem. Note that equations are equivalent if and only if they have the same solution set, e.g.

$$2x+3y = 6 \Leftrightarrow \tfrac{1}{3}x+\tfrac{1}{2}y = 1 \Leftrightarrow y = -\tfrac{2}{3}x+2.$$

An important aim of this chapter is to provide the pupil with quick and efficient techniques for solving systems of linear equations. Four methods are discussed, leaving the teacher to choose which of these to emphasize. The methods are illustrated in the following example:

'Solve the system of equations $3x-2y = 8$, $4x+y = 7$.'

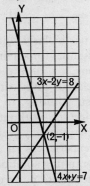

(a) *Graphical* (see Figure). The important point here is the approximate nature of graphical methods, although in combination with 'trial and error' it is possible to find exact solutions in simple cases. The solution set is $\{(2, -1)\}$, which can be verified by substitution.

(b) *Elimination by linear combination.* In this method we use a linear combination technique in which one of the two equations is replaced by an equation involving only one of the variables. In applying this principle to the example chosen,

first write down: $\qquad (3x-2y-8) \qquad (4x+y-7)$

Then fill in appropriate multipliers and complete the sentence:

$$1(3x-2y-8) + 2(4x+y-7) = 0$$

Simplify by using the distributive property:
$$3x-2y-8 + 8x+2y-14 = 0$$
$$\Leftrightarrow 11x-22 = 0$$
$$\Leftrightarrow x = 2, \text{ etc.}$$

(c) *Elimination* (*traditional method*). This method is illuminated by the method of linear combination shown above.

$$\begin{array}{rl} 3x-2y = 8 \times 1 & \Leftrightarrow \quad 3x-2y = 8 \\ 4x+ y = 7 \times 2 & \Leftrightarrow \quad 8x+2y = 14 \\ \text{Add} & \overline{11x \quad\quad = 22} \\ & \Leftrightarrow x = 2, \text{ etc.} \end{array}$$

It is important to write 'add' or 'subtract' to ensure that both sides receive the same treatment. With weaker classes particularly, it may be advisable to choose multipliers which make one coefficient of x (or y) the negative of the other, so that addition is always used.

(d) *The method of substitution.*

Substituting $y = 7-4x$ in $3x-2y = 8$,
$$3x-2(7-4x) = 8$$
$$\Leftrightarrow 3x-14+8x = 8$$
$$\Leftrightarrow 11x = 22, \text{ etc.}$$

This method is particularly suitable when one of the coefficients is 1, and is often essential for solving systems of equations containing non-linear equations.

Section 1. Exercise 1 gives time-consuming experience of the greatest value in substituting, to test whether given ordered pairs belong to the solution set. Without this experience, the abstract work of the rest of the chapter may seem to be a series of tricks.

Section 2. The experiences of Section 1 are now used to give practical ways of graphing solution sets. Since, for example, the line $2x+y = 4$ separates those parts of the plane for which $2x+y < 4$ from those for which $2x+y > 4$, it is evidently important to graph the solution set of the *equation* $2x+y = 4$, then to choose the appropriate half-plane for the inequation. Also included in this Section is essential practice in changing from one form of the linear equation to another.

Section 3 develops the idea of the solution set of a system of inequations in a natural and attractive way. A great deal of use is made of this in linear programming later in the course.

Section 4. See remarks above about the approximate nature of results obtained by graphical methods.

Sections 5 *and* 6. Either of the ways of setting out the solution by elimination may of course be taught as a rote algorithm. Underlying the explanations given is the important idea of linear combination on which the essential notions of linear dependence and independence rest. The following points are worthy of attention:

(i) If two equations in two variables are linearly dependent, their graphs coincide and the solution set is infinite, e.g. $3x - y = 6$ and $3y = 9x - 18$.

(ii) If the parts involving x and y are linearly dependent, but not the whole equations, then the graphs are parallel lines and the solution set is $\emptyset$, e.g. $x + y = 1$, $x + y = 5$.

(iii) If the equations are linearly independent, the graphs intersect and the solution set is a single ordered pair.

Section 7. With beginners it is necessary to emphasize that the notation $\frac{y-3}{5}$ means $\frac{1}{5}(y-3)$, with the line acting as a bracket. It may be helpful to use a combination of line and bracket, i.e. $\frac{(y-3)}{5}$, although it is important not to mislead them into thinking that both brackets and line are needed.

Section 8. Pupils are often able to solve problems mentally through their own natural ingenuity, and the teacher should encourage them in their endeavours. But this inclination should not be allowed to reach a stage where accuracy is endangered. One of the aims in teaching mathematics is to encourage the pupil to explore several different methods of approach to the solution of a problem with a view to selecting the most promising. As this facility develops, a knowledge of mathematical structure builds up also. Some of the problems in the Exercise can be solved mentally. It is useful, nevertheless, to use such problems as a practice ground for the methods under discussion. Some of the questions set in Exercises 8 and 8B pave the way for work to come, such as the coordinate geometry of the straight line and the determination of laws from experimental data.

In solving equations, pupils should be encouraged to check the solutions by substitution. In many cases, this can be done mentally.

(facing page 23)

Systems of Equations and Inequations in Two Variables

1 The solution set of an equation or inequation in two variables

In earlier books we studied the solution sets of open sentences.

Consider the equation $x+2 = 5$, where x is a variable on the set of natural numbers. Replacing x by 3 gives the true sentence $3+2 = 5$, while all other replacements for x give false sentences.

So the solution set of $x+2 = 5$ is $\{3\}$, shown on a number line in Figure 1.

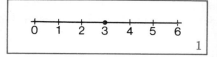
1

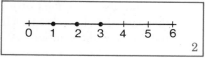
2

In the same way, the solution set of the inequation $x+2 \leqslant 5$, for $x \in N$, is $\{1, 2, 3\}$, shown on a number line in Figure 2.

Suppose we now have *an open sentence in two variables*, such as $2x+y \leqslant 6$, where $x, y \in W$. Since there are two variables, we have to take pairs of replacements for x and y. Here are some pairs which give true sentences:

$x = 0, y = 6; \quad x = 0, y = 0; \quad x = 2, y = 2; \quad x = 3, y = 0.$

We write these as *ordered pairs* (0, 6), (0, 0), (2, 2), (3, 0), and we can illustrate the solution set by regarding the ordered pairs of replacements for x and y as coordinates of points (see Figure 3(i)).

Example 1. Give the solution set of the equation $2x+y = 6$, $x, y \in W$, and show the solution set on the coordinate plane.

The solution set is $\{(0, 6), (1, 4), (2, 2), (3, 0)\}$. See Figure 3(ii).

Example 2. Show the solution set of $2x+y = 6$, $x, y \in R$, on the coordinate plane.

Algebra

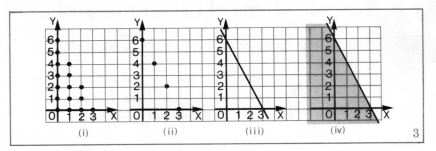

3

As $x, y \in R$ there are infinitely many ordered pairs in the solution set, all lying in the line through the points (0, 6), (1, 4), (2, 2), (3, 0) as shown in Figure 3(iii). The straight line is the *graph* of the solution set of $2x+y = 6$.

Example 3. Indicate graphically the solution set of the inequation $2x+y \leqslant 6$ for $x, y \in R$.

All the points on the line in Figure 3(iv) belong to the solution set, and also all points of a *half plane* on one side of this line, in this case the side which includes the origin. The shaded region is the *graph* of the solution set of $2x+y \leqslant 6$.

Exercise 1

In each of questions *1-4* there is an open sentence and some ordered pairs. Test each ordered pair to see if it belongs to the solution set of the open sentence. Then plot it on 5-mm squared paper, using ● if it does belong, and ○ if it does not belong, to the solution set.

1 $x+y = 4$, $x, y \in W$; (0, 4), (1, 3), (2, 2), (3, 1), (4, 0), (1, 1), (3, 3)

2 $2x+y = 4$, $x, y \in W$; (0, 4), (0, 1), (2, 0), (2, 3), (1, 2)

3 $x+y \leqslant 2$, $x, y \in W$; (0, 0), (0, 1), (0, 2), (0, 3), (3, 0), (2,0), (1, 1), (1, 0)

4 $2x-y = 2$, $x, y \in R$; (0, −2), (0, 2), (2, 2), (2, 1), (3, 4), (3, 2), ($\frac{1}{2}$, −1)

In questions *5-16* find pairs of replacements for x and y which give true sentences, and draw a graph to show (or, in the case of an infinite set, to indicate) the set of points which have these pairs as coordinates.

W, Z, R denote the sets of whole numbers, integers, and real numbers respectively; the variables are on the sets indicated.

Linear equations and inequations in two variables

5 $x+y = 6$; W *6* $x+y = 6$; R *7* $x-y = 0$; W
8 $x-y = 0$; R *9* $2x+y = 8$; W *10* $2x+y = 8$; R
11 $y = 2$; Z *12* $y = 2$; R *13* $x = 4$; Z
14 $x = 4$; R *15* $x+2y \geqslant 4$; R *16* $y \geqslant x$; R

We see from Section 1 that it is often more helpful to show the solution set of an inequation, such as $2x+y \leqslant 6$, graphically than to express it in set-builder form, i.e. $\{(x, y): 2x+y \leqslant 6, x, y \in R\}$.

2 Linear equations and inequations in two variables

The equation $x-2y = 6$ is called a *linear equation* because the graph of its solution set is a straight line. Notice that there is a term in x, a term in y, and a constant term.

In some cases, one of these terms may be missing, as for example in $x = 5$, $x+y = 0$, $3y = 4$, $x = -2y$, all of which are linear equations.

There are several equivalent ways of writing a linear equation. For example, $\quad 3x+2y+6 = 0 \quad$ (in the form $ax+by+c = 0$)
$$\Leftrightarrow 3x+2y = -6$$
$$\Leftrightarrow y = -\tfrac{3}{2}x-3 \quad \text{(in the form } y = mx+c\text{)}$$

Example 1. Indicate graphically the solution set of $x-2y+6 = 0$, $x, y \in R$.

We know that the graph of the solution set will be a straight line. It crosses the x-axis where $y = 0$, and $x = -6$; i.e. at $(-6, 0)$. It crosses the y-axis where $x = 0$, and $y = 3$; i.e. at $(0, 3)$.

Alternatively. Put the equation in the form $y = \tfrac{1}{2}x+3$, and compare with the equation of a straight line $y = mx+c$, which has gradient m and cuts the y-axis at $(0, c)$, as shown in Book 3 Geometry, Chapter 3. Here the gradient is $\tfrac{1}{2}$ and the graph cuts the y-axis at $(0, 3)$.

The graph of the solution set is shown in Figure 4.

Example 2. Indicate graphically the solution set of $x+2y \geqslant 4$, $x, y \in R$.

Algebra

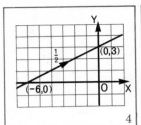
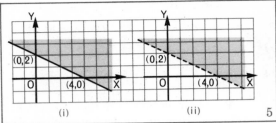

4

(i)   (ii)

5

The solution set of the *equation* $x+2y = 4$ is indicated by the straight line through the points $(4, 0)$ and $(0, 2)$.

All points of the solution set of $x+2y > 4$ lie in one of the half planes bounded by this line. Test $(0, 0)$. Since this is not a solution, the half plane required is not the one containing the origin.

The solution set of $x+2y \geqslant 4$ is the *union* of the two sets forming the line and the shaded half plane shown in Figure 5(i).

Note: The line is *not* included in the graph of the solution set of $x+2y > 4$, and is then shown as a broken line, as in Figure 5(ii).

Exercise 2

In questions *1-12* express the given equations in each of the forms:

a   'x-term + y-term + constant = zero'   e.g. $x-2y+6 = 0$
b   'x-term + y-term = constant'   e.g. $x-2y = -6$
c   '$y = x$-term + constant'   e.g. $y = \frac{1}{2}x+3$

1    $2x+y-4 = 0$ 2 $2x-y-1 = 0$ 3 $x+3y-6 = 0$

4    $x+2y = 3$ 5 $x-y = 5$ 6 $2x-y = -1$

7    $2x-3y-6 = 0$ 8 $3x+2y+6 = 0$ 9 $y = \frac{3}{4}x-3$

10    $y = \frac{1}{2}x+1$ 11 $x = -2$ 12 $y = 3$

Indicate the solution sets of the following by sketches on plain or squared paper; the variables are on the set of real numbers.

13    $x-3y = 6$ 14 $x-3y \leqslant 6$ 15 $x-3y < 6$

16    $2x+y = 4$ 17 $2x+y \geqslant 4$ 18 $2x+y > 4$

19    $2x-3y-6 = 0$ 20 $2x-3y-6 \geqslant 0$ 21 $2x-3y-6 < 0$

The solution set of a system of inequations

22 $x \geq -2$ 23 $y < 3$ 24 $x \leq 0$
25 $3x - 2y = -6$ 26 $3x - 4y = 12$ 27 $x + 4y \leq 4$
28 $x + y \geq 5$ 29 $x - y < 8$ 30 $4x + 3y - 12 > 0$

3 The solution set of a system of inequations (with variables on R)

For the rest of this chapter we will assume that the variables are on the set of real numbers (R).

In the study of Linear Programming later in the course we will often be interested in the solution set of a *system of inequations*, for example:

$$x \geq 0 \text{ and } y \geq 0 \text{ and } x + 2y \leq 4.$$

Compare this with the system of open sentences:

'He studies mathematics *and* he has blue eyes *and* he is in the football team.' We saw that the solution set of this kind of system is the *intersection* of the three open sentences, as shown in the Venn diagram in Figure 6.

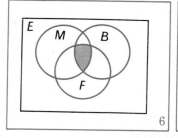

6

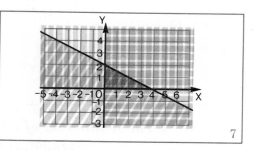

7

In exactly the same way, the solution set of the above system of inequations is the *intersection* of the solution sets of the three inequations, as shown in Figure 7.

Exercise 3

Show by shading, or by the use of colour, the solution sets of the systems of inequations in questions *1-6*.

Algebra

1. $x \geq 0$ and $y \geq 0$ and $x+y \leq 5$
2. $x \geq 0$ and $y \geq 0$ and $2x+y \leq 6$
3. $x > 0$ and $y > 0$ and $x+y < 8$
4. $x > 0$ and $y > 0$ and $2x+3y < 12$
5. $x \leq 0$ and $y \leq 0$ and $x+y \geq -6$
6. $x > 0$ and $y < 0$ and $x+2y > 8$
7. Explain why the solution set of $x \geq 3$ and $x < 2$ is empty.
8. Show the solution set of $-2 \leq x$ and $x \leq 2$. The notation $-2 \leq x \leq 2$ is often used for this *system* of inequations.
9. How many inequations are there in this system? $-2 \leq x \leq 2$ and $-4 \leq y \leq 4$. Illustrate their solution set.

In the following questions shade or colour the separate solution sets lightly, and indicate clearly the intersection of the solution sets. Notice that commas often replace 'and' in systems of equations and inequations.

10. $y \geq 2, x+y \leq 6$
11. $x \geq 0, y \leq x$
12. $y \leq 0, y \leq x$
13. $y \leq 6, y \geq x, y \geq -x$
14. $x < 10, y < x, y > -x$
15. $y \geq 0, y \leq x, x+y \leq 5$
16. $x \geq 0, y \geq 0, y \leq 5, y \leq 8-x$
17. $x+y \leq 2, y \geq x-4$
18. $y > 2x-1, x+2y \geq 6, y \leq 5$

4 The solution set of a system of equations —graphical method

The two straight lines in Figure 8 indicate the solution sets of the equations $x+y = 5$ and $x-y = 1$. The solution set of the system of equations $x+y = 5$ *and* $x-y = 1$ is the intersection of the two solution sets, and is given by the coordinates of the point where the two lines cross. This point is A (3, 2), so the solution set of the system is $\{(3, 2)\}$.

The solution set of a system of equations—graphical method

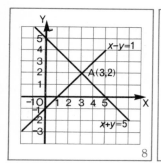

8

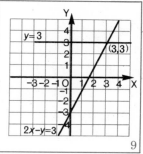

9

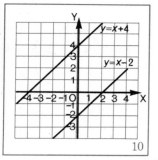

10

Example 1. Find graphically the solution set of the system of equations $y = 3$ and $2x - y = 3$.

Figure 9 shows the solution set of the two equations. The solution set of the system is the intersection of these two solution sets, i.e. $\{(3, 3)\}$.

Example 2. Explain with the aid of a graph why the solution set of the system $y = x + 4$, $y = x - 2$ is the empty set.

Figure 10 shows that the lines indicating the solution sets of the two equations are parallel. They do not intersect, therefore, and the solution set of the system of equations is ø.

Exercise 4

Use 5-mm squared paper to find graphically the solution set of each system of equations in questions *1-8*.

1 $x = 3$ and $y = 4$ 　　　　*2* $x = 0$ and $y = -2$
3 $x + y = 7$ and $y = 3$ 　　*4* $x + y = 6$ and $x = -3$
5 $x + y = 8$ and $y = x$ 　　*6* $y = x + 2$ and $y = 4 - x$
7 $y = x + 5$ and $y = x - 5$ 　*8* $x - 2y = 3$ and $x + y = 0$

Use 2-mm squared paper to find graphically the solution set of each system of equations in questions *9-20*, rounded to 1 decimal place. Take a scale of 2 cm (10 squares) to represent 1 unit on each axis, and first find the points where the lines cut the axes so that you can decide what length to make the axes.

Algebra

9 $y = x, x+y = 1$
10 $x+y = 2, 2x-y = 2$
11 $3x+2y = 6, x-y = 1$
12 $2x+y+2 = 0, y = x+2$
13 $x-y = 1, 3x+4y = 12$
14 $x+3y = 3, x+y = 0$
15 $x-y = -4, x+2y = 4$
16 $3x-2y+6 = 0, 3x-2y = 0$
17 $x+y = 3, x-y = 2$
18 $5x+3y = 7, 3x-5y = 0$
19 $x+3y = 9, x+y = 5, 2x-y = 4$
20 $x+y = 2, 3x-2y = 11, 2x-y = 7$

5 The solution set of a system of equations —elimination method

Figure 11 shows the graphs of the equations $x+y-6 = 0$ and $2x-y = 0$. It also shows the graphs of $p(x+y-6)+q(2x-y) = 0$ for several different replacements of p and q.

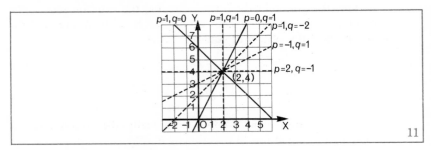

Notice that all the lines pass through the same point. It is easy to see why this is so. Replacing x and y by the coordinates of this point make both $x+y-6 = 0$ and $2x-y = 0$ into true sentences. Therefore the same replacements will make $p(x+y-6)+q(2x-y) = 0$ into a true sentence whatever p and q may be, since $(p \times 0)+(q \times 0) = 0$.

Whichever two (or more) of these lines we choose, we will obtain equations giving a system *equivalent* to the original system, i.e. with the same solution set. In this case the most convenient pair of lines to choose is $x = 2$ and $y = 4$, giving the solution set immediately as $\{(2, 4)\}$.

The solution set of a system of equations—elimination method

Example 1. Find the solution set of the system $x+y = 8, x-y = 2$
First rewrite them in the form $x+y-8 = 0, x-y-2 = 0$.
Then consider $p(x+y-8)+q(x-y-2) = 0$.

There are many possible choices for p and q which will make either the coefficient of x or y zero, so eliminating x or y.

One choice is $p = 1, q = 1$, giving $(x+y-8)+(x-y-2) = 0$
$$\text{Then} \quad 2x-10 = 0$$
$$\text{i.e.} \quad x = 5$$

The point on the line $x+y = 8$ for which $x = 5$ has y-coordinate given by $5+y = 8$, or $y = 3$.

The solution set is $\{(5, 3)\}$. See Figure 12.

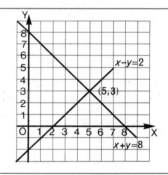

12

Example 2. Solve the system $\quad 3x-2y = 8, 4x+y = 7$
First stage: $\quad\quad\quad 3x-2y-8 = 0, \quad 4x+y-7 = 0$
Second stage: $\quad\quad 1\,(3x-2y-8) +2\,(4x+y-7) = 0$
Multiplying out: $\quad\quad\quad\quad\quad\quad\quad\quad 11x-22 = 0$
$$x = 2$$

From $4x+y = 7, 8+y = 7$, so $y = -1$
The solution set is $\{(2, -1)\}$.

Note: The working may be set out as follows.

Example 1. Find the solution set of the system $x+y = 8$, $x-y = 2$.

$$x+y = 8$$
$$x-y = 2$$

Add $\quad\quad 2x \quad = 10$

$\Leftrightarrow \quad x \quad = 5$

$$x+y = 8$$
$$x-y = 2$$

Subtract $\quad\quad 2y = 6$

$\Leftrightarrow \quad y = 3$

The solution set is $\{(5, 3)\}$.

Algebra

Example 2. Solve the system of equations $3x-2y = 8$,
$4x+y = 7$

$$\begin{aligned} 3x-2y = 8 \quad &\times 1 \quad \Leftrightarrow \quad 3x-2y = 8 \\ 4x+y = 7 \quad &\times 2 \quad \Leftrightarrow \quad 8x+2y = 14 \\ &\text{Add} \quad 11x \quad\quad = 22 \\ &\Leftrightarrow \quad x \quad\quad\quad = 2 \end{aligned}$$

Substituting $x = 2$ in the second equation,

$$\begin{aligned} (4\times 2)+y &= 7 \\ \Leftrightarrow \quad 8+y &= 7 \\ \Leftrightarrow \quad y &= -1 \end{aligned}$$

The solution set is $\{(2, -1)\}$.

Exercise 5

Solve each of the following systems of equations by the method of elimination.

1. $x+y = 12,\ x-y = 4$ 2. $x+y = 6,\ x-y = 0$
3. $x+y = 5,\ x-y = 2$ 4. $x+2y = 3,\ -x+3y = 2$
5. $2x-y = 4,\ -2x-3y = -4$ 6. $-3x+4y = 7,\ 3x+y = -2$
7. $2x+3y = 5,\ x+y = 2$ 8. $2x+5y = 16,\ x-y = 1$
9. $2x-y = 5,\ x-2y = 4$ 10. $3x+4y = -7,\ 2x+y = -3$
11. $x+y = 0,\ 4x+3y = 3$ 12. $2x-5y = 1,\ 4x-3y = 9$
13. $2x+4y = 7,\ 4x-3y = 3$ 14. $4x-5y = 22,\ 7x+3y = 15$
15. $2y-3x = 0,\ x-y+2 = 0$ 16. $2x+3y-8 = 0,\ 3x+2y-17 = 0$
17. $7x+4y-1 = 0,\ 5x+2y+1 = 0$
18. $2x-7y+3 = 0,\ 3x-7y+1 = 0$
19. $5x = 8y,\ 4x-3y+17 = 0$ 20. $x+y = 2,\ 2x-y = 2$
21. $3x+2y-6 = 0,\ x-y-1 = 0$ 22. $2x+y+2 = 0,\ x-y+2 = 0$
23. $3x = 4y+12,\ y = x-1$ 24. $4x = 5y,\ 3y = 7-5x$
25. $3x-5y = 2,\ 7x+3y = 12$ 26. $11x+3y+7 = 0,\ 2x+5y-21 = 0$

27 $7x+3y-15 = 0$, $5x-2y = 19$
28 $5x-2y = 0·6$, $2x+y = 1·5$

6 The solution set of a system of equations —substitution method

Substitution means 'putting in place of'. The method of substitution is a convenient way to solve a system of equations if one of the terms in x or y has coefficient 1.

Example. Solve the pair of equations $3x-2y = 8$, $4x+y = 7$.
The *second* equation can be written $y = 7-4x$.
Put $(7-4x)$ in place of y in the *first* equation:

$3x-2(7-4x) = 8$ Substituting $x = 2$ in the first equation,
⇔ $3x-14+8x = 8$ $6-2y = 8$
⇔ $11x = 22$ ⇔ $-2y = 2$
⇔ $x = 2$ ⇔ $y = -1$

The solution set is $\{(2, -1)\}$.

Exercise 6

Solve each of the following systems of equations by the method of substitution.

1   $y = x$, $2x-y = 5$ 2 $y = 2x$, $6x-y = 8$
3   $x = 3y$, $2x-3y = 12$ 4 $x = 2y$, $3x-10y = 12$
5   $x = y+1$, $x+5y = 7$ 6 $x = y+4$, $x+3y = 16$
7   $y = x+3$, $2x+3y = 4$ 8 $y = x-5$, $x+2y = 20$
9   $y = 2x-1$, $3y-2x = 5$ 10 $x = 2y+9$, $x+5y+5 = 0$
11 $u = 2v+3$, $5u-2v+1 = 0$ 12 $q = 2p-2$, $5p-4q+1 = 0$
13 $a-2b = 2$, $3b-5a = 4$ 14 $y+2x = 0$, $4x+y = 3$
15 $2x+y = 0$, $7x+5y = 1$ 16 $x-2y+1 = 0$, $2x+3y = 12$
17 $6u-v = 1$, $4u-3v+4 = 0$ 18 $a+b+1 = 0$, $9a+8b+7 = 0$
19 $5p+q = 10$, $14p+3q = 18$ 20 $s-8t+20 = 0$, $5s-7t+1 = 0$

7 Systems of equations with fractions

When one or both equations contain fractions it is usually best to form equivalent equations without fractions, and then to use the methods of Sections 5 and 6.

Example. Solve $x - y = -1$, $\dfrac{x}{2} - \dfrac{y-3}{5} = 1$

$$\dfrac{x}{2} - \dfrac{y-3}{5} = 1$$

$$\begin{aligned}
&\Leftrightarrow\quad \tfrac{1}{2}x - \tfrac{1}{5}(y-3) = 1 \\
&\Leftrightarrow\quad \tfrac{10}{2}x - \tfrac{10}{5}(y-3) = 10 \quad \text{(multiplying each side by 10,} \\
&\Leftrightarrow\quad 5x - 2(y-3) = 10 \quad \text{the LCM of 2 and 5)} \\
&\Leftrightarrow\quad 5x - 2y + 6 = 10 \\
&\Leftrightarrow\quad 5x - 2y = 4
\end{aligned}$$

The pair of equations $x - y = -1$ and $5x - 2y = 4$ can now be solved as before. Show that the solution set is $\{(2, 3)\}$.

Exercise 7

Solve each of the following systems of equations.

1. $2x - 3y = 5$, $\tfrac{1}{2}x - y = 1$
2. $x - y = 5$, $\tfrac{1}{5}x + y = 1$
3. $\tfrac{2}{3}x + y = 1$, $x + y = 2$
4. $\tfrac{1}{4}x - \tfrac{1}{2}y = 0$, $x + y = 6$
5. $x - y = 0$, $\tfrac{1}{2}x + \tfrac{1}{3}y = 5$
6. $2x - y = 7$, $\tfrac{3}{4}x - \tfrac{1}{2}y = 3$
7. $x + \tfrac{1}{2}y = 4$, $\tfrac{1}{3}x - y = -1$
8. $3x - \tfrac{1}{2}y = 5$, $\tfrac{1}{3}x + \tfrac{1}{4}y = 3$
9. $\tfrac{1}{3}x + 5 = \tfrac{2}{3}y$, $\tfrac{1}{2}x + \tfrac{1}{3}y = \tfrac{1}{2}$
10. $\tfrac{1}{2}x + \tfrac{1}{5}y = 1$, $\tfrac{1}{3}x - \tfrac{1}{5}y = 4$

Exercise 7B

Solve each of the following systems of equations.

1. $x - y = 0$, $\tfrac{1}{4}x - \tfrac{1}{3}(y - 1) = 0$
2. $\tfrac{1}{2}x - \tfrac{1}{3}(y + 6) = 2$, $2x + 3y = 3$
3. $2x + 5y = 4$, $\tfrac{1}{2}(x + 1) + \tfrac{1}{3}(y - 2) = -1$
4. $5x - 2y = 11$, $\tfrac{1}{2}(x - 1) - \tfrac{1}{5}(y - 3) = 1$

Systems of equations in two variables

5    $\dfrac{2x+y}{3} = 5, \dfrac{3x-y}{5} = 1$ 6 $\dfrac{x+1}{3} + y = 8, \; x - \dfrac{y+1}{3} = -4$

7    $5x + 3y = 2x + 7y = 29$ 8 $\dfrac{x-1}{6} + y = 6, \; \dfrac{y-1}{4} + x = 8$

9    $\dfrac{2x-3y}{4} = \dfrac{3x-2y}{5} = 7\tfrac{1}{2}$ 10 $0.3x + 0.5y = 4.7,$
                                                                          $0.9x - 0.2y = 2.2$

Solve the following systems by first substituting p for $\dfrac{1}{x}$ and q for $\dfrac{1}{y}$.

11    $\dfrac{2}{x} + \dfrac{3}{y} = 2, \; \dfrac{8}{x} - \dfrac{9}{y} = 1$ 12 $\dfrac{1}{x} + \dfrac{1}{y} = \dfrac{3}{4}, \; \dfrac{2}{x} + \dfrac{3}{y} = \dfrac{5}{2}$

8 Problems leading to systems of equations in two variables

In the problems in this section, as in many problems, the appropriate mathematical model is a system of linear equations.

If you cannot see the solution at once, analyse the problem carefully, and construct a mathematical model in the form of a system of equations. Then find the solution set for this system, and interpret it in terms of the original problem.

Example. A rectangle is to be drawn with perimeter 75 cm, and the length is to be 13 cm more than the breadth. What should the length be?

Take x cm to be the length, and y cm to be the breadth. Then $x = y + 13$ *and* $2x + 2y = 75$ gives a suitable mathematical model. Solving this by substitution:

             $2(y+13) + 2y = 75$
⇔ $2y + 26 + 2y = 75$
⇔ $4y = 49$
⇔ $y = 12\tfrac{1}{4}$ $x = 12\tfrac{1}{4} + 13 = 25\tfrac{1}{4}$

The length should be 25.25 cm.

Notice the stages—introducing the variables, forming the equations, solving the system, giving the solution of the original problem.

Algebra

Exercise 8

In the problems below, first see if you can solve them by inspection. Then, whether you can or not, use the method explained above to solve them.

1. The sum of two whole numbers is 112, and their difference is 36. Find the numbers.

2. The sum of two rational numbers is 63, and their difference is 12. Find the numbers.

3. Two angles are complementary. One is 57° larger than the other. Find the sizes of the angles.

4. Two angles are supplementary. One is 68° larger than the other. Find the sizes of the angles.

5. The sum of the length and breadth of a rectangle is 84 cm. The length is 18 cm more than the breadth. Calculate the length and breadth.

6. The perimeter of a rectangle is 68 cm. The length is 27 cm more than the breadth. Find the length and breadth.

7. Six records of one make and four of another make cost £3·40. Three of the first kind and ten of the second kind cost £4·90. Find the price of each type of record.

8. Twelve expensive flower bulbs and eight cheap ones cost £3·80. Nine of the expensive ones and four of the cheap ones cost £2·65. Find the price of each kind of bulb.

9. A straight line has equation $y = mx + c$. (2, 2) and (3, 6) are points on the line. Form a pair of equations, and solve them to find m and c. If the point $(a, 14)$ lies on the line, find a.

10. Repeat question *9* if the points $(-1, -2)$ and $(3, 10)$ lie on the line with equation $y = mx + c$. Does the point $(-2, -7)$ lie on the line?

Systems of equations in two variables

Exercise 8B

1. $ax+by = 29$ when $x = 3$ and $y = 5$, and $ax+by = 13$ when $x = 7$ and $y = -2$. Find a and b.

2. The straight line $px+qy = 1$ passes through the points $(3, -2)$ and $(-5, 4)$. Find p and q.

3. For the series $3+9+15+21+\ldots$, the nth term is of the form $nd+k$, where d and k are constants. Using the first and second terms of the series, write down two equations in d and k. Hence find d and k. Apply your results to calculate the 100th term of the series.

4. The speed V metres per second of a train t seconds after the brakes are applied is given by $V = at+b$, where a and b are numbers. When $t = 0$, $V = 16$; when $t = 8$, $V = 10$. Find a and b.

   a Calculate V when $t = 10$.
   b Find t at the instant when the train stops.

5. The height h metres above the ground reached by a projectile after t seconds is given by formula $h = at+bt^2$. Find the constants a and b given that $h = 19$ when $t = 1$, and $h = 28$ when $t = 2$. Use the formula to calculate h when $t = 4$. What happens when $t = 4.8$?

6. In the equation $p = \dfrac{x^2}{ax+b}$, $p = -1$ when $x = 1$, and $p = 3$ when $x = 3$. Calculate a and b.

7. The ratio of two positive integers x and y is $5:3$, and their difference is 56. Find the two integers.

8. A motorist does a journey of x km in t hours at an average speed of 68 km per hour. Write down an equation in x and t.
   If he could reduce his time by 10 minutes, his average speed would be 72 km per hour. Write down another equation in x and t. Hence find:

   a the time taken for the journey   b the distance travelled.

9. If $x+2y = p$ and $x-y = q$, show that $y = \frac{1}{3}(p-q)$ and find a corresponding equation for x by eliminating y.

Algebra

10 If $2x+y = a+3b$ and $x-2y = 3a-b$, find x and y in terms of a and b.

11 480 people attended a Charities concert. Front stalls cost 40p each and back stalls cost 60p each. The total receipts were £253. How many front stalls and how many back stalls were occupied?

12 An engineering firm has a machine X which turns out 30 components per hour and has installed a new machine Y which turns out 40 components of the same kind per hour. If 600 of these components were produced on a particular day when the total number of hours of machine operation was 18, for how many hours were X and Y operated? (Suppose that X and Y were operated for x hours and y hours respectively.)

Summary

1 A *linear equation* can be written in several equivalent forms:

   (i) $ax + by + c = 0$ (ii) $ax + by = -c$ (iii) $y = mx + c$

2 The *graph of a linear equation* is a straight line which can be drawn:

   (i) by finding the coordinates of the points where it cuts the x-axis and y-axis
   (ii) by writing the equation in the form $y = mx + c$ which gives the point $(0, c)$ on the y-axis and the gradient m.

3 The *graph of a linear equation*, e.g. $2x - y + 2 = 0$, divides the coordinate plane into two half planes each of which illustrates the solution set of an inequation (see Figure 1).

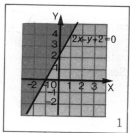
1

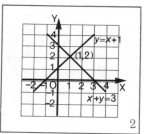
2

4 The *solution set of a system of equations or inequations* is the *intersection* of the solution sets of the separate equations or inequations.

   a In the case of systems of inequations, the solution set is best given graphically.
   b In the case of systems of equations, the solution set can be found:

   (i) approximately, *by graphing the solution sets* to find the point of intersection (see Figure 2)
   (ii) by *eliminating one of the variables*
   (iii) by *substitution*.

5 Some problems may be solved by setting up a suitable mathematical model made up of a system of equations or inequations.

Formulae and Literal Equations

1 Constructing and using a formula

We have already constructed and used several formulae. Do you recognize the following?

$A = lb$, $A = l^2$, $A = \frac{1}{2}bh$, $A = \pi r^2$, $D = ST$,
$P = 2(l+b)$, $C = 2\pi r$, $V = lbh$, $C = \frac{5}{9}(F-32)$.

In this chapter we study formulae more closely, and consider various ways in which they are used.

(i) Bus routes

Figure 1 shows bus routes between three towns. How many routes are there from Ayton to Beeton? For each of these routes, how many routes are there from Beeton to Seaton? So how many routes could you possibly take from Ayton to Seaton?

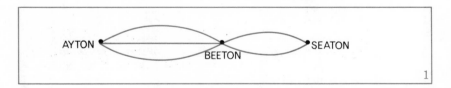

1

If there were m routes from Ayton to Beeton and n routes from Beeton to Seaton, we could argue like this:

There are m routes from Ayton to Beeton; whichever of these we take, we have the choice of n routes from Beeton to Seaton. Then if R denotes the number of routes possible from Ayton to Seaton we have the formula

$$R = m \times n$$

If there were p routes from Seaton to a fourth town, Deeton, what would be the formula for the number (R') of different routes from Ayton to Deeton?

Note to the Teacher on Chapter 3

Section 1 extends the scope of constructing and using a formula given in Book 2, and many of the questions in Exercise 1 are largely revision. The analysis of a situation and the construction of a mathematical model—in this instance a formula—is often as important as, and may be more useful than, actually using the formula. It should be noted that while, in Example (ii), page 41, the area of the plate is A m², the formula connecting a, b, and A is $A = \frac{3}{2}ab$.

As a preliminary to the study of Literal Equations in *Section* 3, *Section* 2 summarizes and revises work on equations in one variable; further practice may be given here if necessary.

In *Section* 3, the parallel study of equations with literal and numerical constants given on page 48 is considered to be very important; if the pupil understands that x is a variable and that a, b, c, etc., represent specific numbers, and realizes that the techniques used in obtaining the solutions are the same, much difficulty should be avoided.

Section 4 introduces cross-multiplication, a technique that will be useful in several parts of the course; for example, in solving equations of the kind shown in the two worked examples, in calculations involving the lengths of the sides of two similar triangles, and in the use of the Sine Rule in trigonometry. Teachers should emphasise that cross-multiplication can only be used when there are two equal ratios in the form $\frac{x}{a} = \frac{b}{c}$.

The work in *Section* 5 on Changing the Subject of a Formula follows from that in *Section* 3. Note that in Exercise 5 most of the questions require the transformation of a formula followed by an application of the formula in its new form, in order to provide some motivation.

Many pupils in the past have found the ideas and the manipulation involved in this chapter difficult. It is hoped that the progressive steps taken through the various sections, and the carefully chosen questions in Exercises 1, 2, 3, 4 and 5, will give most pupils some confidence and success, and that the harder questions in Exercises 1B, 3B, 4B and 5B will provide an appropriate challenge and sufficient practice for the abler pupils.

Constructing and using a formula

The important thing about a formula is that, once established, it can be used in any calculation concerning the situation to which it refers.

For example, if there were 5 routes from Ayton to Beeton and 4 from Beeton to Seaton, then $m = 5$, $n = 4$, and $R = m \times n$
$$= 5 \times 4$$
$$= 20$$
so there would be 20 routes from Ayton to Seaton.

Use the formula $R' = mnp$ to find the number of routes from Ayton to Deeton, given that there are 4 from Ayton to Beeton, 3 from Beeton to Seaton, 2 from Seaton to Deeton.

(ii) The area of a metal plate

Figure 2 represents a metal plate in the shape of a rectangle and an isosceles triangle with the dimensions shown.

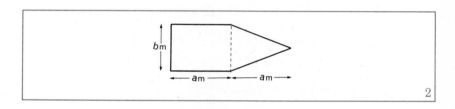

Denote the area of the whole plate by A m², the area of the rectangle by R m², and of the triangle by T m².

$$R = ab, \text{ and } T = \tfrac{1}{2}ab$$
$$A = R+T$$
$$= ab + \tfrac{1}{2}ab$$
$$= \tfrac{3}{2}ab$$

When $a = 10$ and $b = 7$, $A = \tfrac{3}{2} \times 10 \times 7 = 105$.
What is the area when $a = 8$ and $b = 5$?

(iii) The sum of the angles of a polygon with n sides

A polygon can be divided into triangles as indicated in Figure 3. What is the sum of the angles of a triangle in degrees and in right angles?

Algebra

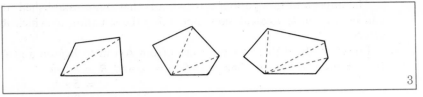

Copy and complete this table:

Number of sides	4	5	6	7	8	10	20	n
Number of triangles	2	3						
Sum of angles of polygon in right angles	4							

Suppose the sum of the angles of a polygon with n sides is S right angles; give the formula for S in terms of n.

For a polygon with 100 sides, what is the sum of the angles:

a in right angles *b* in degrees?

Exercise 1

1 A rectangle is x cm long and y cm broad.

  a Write down formulae for the perimeter P cm and the area A cm^2 of the rectangle.
  b Calculate the perimeter and area when $x = 12$ and $y = 8$.

2 The sides of a square are s cm long.

  a Write down formulae for the perimeter P cm and the area A cm^2 of the square.
  b Calculate the perimeter and area when $s = 15$.

3 Figure 4 shows the route of an aircraft which flew x km east, then y km north. The total distance it flew from O to E to N was d km.

  a Write down a formula for d.
  b Hence calculate d if $x = 840$ and $y = 460$.

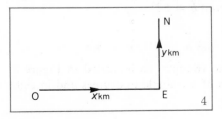
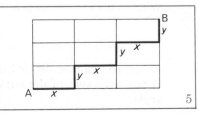

Constructing and using a formula

4. Figure 5 shows a network of streets in a town. All the units are metres.
   a. Find a formula for the distance d, along the streets shown, from A to B.
   b. Calculate d if $x = 325$ and $y = 180$.

5. A lorry weighs x tonnes when empty. If y tonnes of coal are added, its loaded weight is W tonnes. Write down a formula for W.
   a. Find W when $x = 2 \cdot 75$ and $y = 2 \cdot 50$.
   b. Find y when $x = 3\frac{1}{2}$ and $W = 6$.

6. A plank of wood is p metres long. When q metres have been sawn off, the remaining length is L metres. Write down a formula for L, and hence find L when $p = 5$ and $q = 1 \cdot 5$.

7. For Figure 6 write down a formula for perimeter P in terms of x, y, and z.
   a. Find P when $x = 35$, $y = 15$, and $z = 12$.
   b. Find z when $P = 64$, $x = 12$, and $y = 10$.
   c. Find x when $P = 36$, $y = 5$, and $z = 12$.

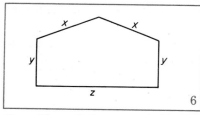
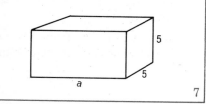

8. Figure 7 shows a cuboid with length a mm, breadth 5 mm and height 5 mm.
   a. Find a formula for the sum S mm of the lengths of all the edges of the cuboid.
   b. Calculate S if $a = 15$.

9. a. $A = \{a\}$. List the subsets of A. (Do not forget the empty set.) How many are there?
   b. $B = \{a, b\}$. List the subsets of B. How many are there?
   c. $C = \{a, b, c\}$. List the subsets of C. How many are there?
   d. Can you guess the number of subsets of $D = \{a, b, c, d\}$?
   e. Can you suggest a formula for the number of subsets there are for a set containing n elements? (*Hint*: Express your answers as powers of 2.)

Algebra

10 In this question you have to obtain a formula for the greatest number of diagonals that can be drawn from one chosen vertex in a polygon.

a Refer back to Figure 3, page 42, and copy and complete this table:

Number of sides	4	5	6	7	8	10	20	n
Number of diagonals	1	2						

b If the greatest number of diagonals that can be drawn from any chosen vertex in a polygon with n sides is d, give a formula for d in terms of n.

c How many diagonals can be drawn from one vertex of a polygon with 100 sides?

Exercise 1B

1 The two sets of equally spaced parallel lines in Figure 8 represent a network of roads in a town. The distances between adjacent junctions are x metres and y metres as shown. If the shortest distance by road from A to B is denoted by b metres, from A to C by c metres, from A to D by d metres, and so on, we could write a formula for the distance from A to C: $c = 3x+y$. Write down formulae for the distances:

a from A to B   *b* from A to D   *c* from A to E   *d* from A to F.
Calculate each of these distances when $x = 75$ and $y = 50$.

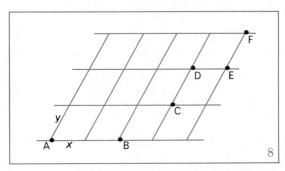
8

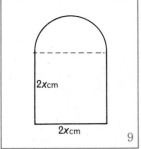
9

2 Figure 9 represents a metal plate in the form of a square of side $2x$ cm and a semicircle. Write down formulae for:

a the perimeter P cm of the plate   *b* the area A cm^2 of the plate.
Calculate P and A when $x = 10$ and 3·14 is taken as an approximation for π.

Constructing and using a formula

3 *a* Find possible 5th, 6th, and 7th numbers of the sequence 3, 7, 11, 15, ...
  b Can you write down the 100th number?
  c The nth number N is given by the formula $N = 4n-1$. Check that this formula gives the 5th, 6th, and 7th numbers, and use it to find the 100th and the 1000th numbers.

4 Figure 10 shows a trellis of squares, each of side 1 cm.
  a How many squares of side 1 cm are there?
  b How many points are there where two lines cross or meet?
  c If there were x squares across and y squares up,
    (*1*) find a formula for the number S of squares of side 1 cm,
    (*2*) find a formula for the number P of points where two lines cross or meet, and
    (*3*) find S and P for the case when $x = 29$ and $y = 30$.

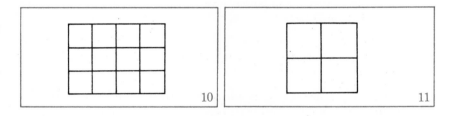

5 In Figure 11 how many squares are there altogether in the 2×2 square tiling shown? (Not four!)
  a Draw a 3×3 tiling. How many squares are in it? (Not nine or ten!)
  b For an $a \times a$ tiling, the number of squares N is given by the formula $N = \frac{1}{6}a(a+1)(2a+1)$. Use this formula to check your answers for the 2×2 and 3×3 tilings.
  c Calculate the number of squares in a 100×100 tiling.

6 Figure 12 shows a quadrilateral (four sides) and a pentagon (five sides), with all the diagonals drawn.
  a Write down the total number of diagonals in each figure.
  b Draw a hexagon (six sides) with all its diagonals. How many diagonals are there?
  c For a figure with n sides (a polygon), the number of diagonals N is given by the formula $N = \frac{1}{2}n(n-3)$.
  Check your answers to *a* and *b* by means of this formula.

Algebra

d How many diagonals would a figure of twenty sides have?
e How many sides would a polygon with 54 diagonals have?

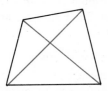 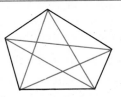

12

2 Revising the methods of solving equations

In the algebra course so far we have solved equations in one variable of the following kinds:

$2x = 8$, $x+2 = 9$, $\frac{1}{4}x = 10$ . . . . in Book 1
$2x+4 = -9$, $4(1-2x) = 12-7x$ . . . in Book 2
$\frac{1}{2}(x-2) - \frac{1}{3}(x-3) = 1$ . . . . . in Book 3

You will remember that all these equations could be solved by adding appropriate negatives to each side, and by multiplying each side by appropriate reciprocals. The following examples illustrate the methods.

Revision examples

1
$\quad\quad 6x+12 = -9$
$\Leftrightarrow 6x+12-12 = -9-12$
$\Leftrightarrow \quad\quad 6x = -21$
$\Leftrightarrow \quad\quad \frac{1}{6} \times 6x = \frac{1}{6} \times (-21)$
$\Leftrightarrow \quad\quad\quad x = -\frac{7}{2}$
The solution set is $\{-\frac{7}{2}\}$.

2
$\quad\quad \frac{3}{4}x = 10$
$\Leftrightarrow \frac{4}{3} \times \frac{3}{4}x = \frac{4}{3} \times 10$
$\Leftrightarrow \quad x = \frac{40}{3}$, or $13\frac{1}{3}$
The solution set is $\{\frac{40}{3}\}$ or $\{13\frac{1}{3}\}$.

3
$\quad\quad 4(1-2x) = 12-7x$
$\Leftrightarrow \quad 4-8x = 12-7x$
$\Leftrightarrow \quad 7x-8x = 12-4$
$\Leftrightarrow \quad\quad -x = 8$
$\Leftrightarrow \quad\quad\quad x = -8$
The solution set is $\{-8\}$.

4
$\quad\quad \frac{1}{2}x + \frac{1}{3}x = -2$
$\Leftrightarrow 6(\frac{1}{2}x + \frac{1}{3}x) = 6 \times (-2)$ [LCM is 6]
$\Leftrightarrow \quad 3x + 2x = -12$
$\Leftrightarrow \quad\quad 5x = -12$
$\Leftrightarrow \quad\quad x = -\frac{12}{5}$
The solution set is $\{-\frac{12}{5}\}$.

Literal equations

Exercise 2

Solve the following equations, where x is a variable on the set of real numbers.

1. $2x = 5$
2. $3x = 8$
3. $2x = -1$
4. $5x = -4$
5. $x - 3 = 4$
6. $x - 2 = -1$
7. $x + 3 = 0$
8. $x + 4 = -7$
9. $3x - 1 = 8$
10. $2x - 5 = 0$
11. $4x + 1 = 6$
12. $6x - 6 = -3$
13. $\frac{1}{2}x = 4$
14. $\frac{1}{3}x = -1$
15. $\frac{2}{3}x = 5$
16. $\frac{3}{4}x = -9$
17. $2(x+1) = 3$
18. $3(x-1) = -2$
19. $5(x-2) = 0$
20. $2(1-x) = 1$
21. $6(2x+1) = 0$
22. $4(1-2x) = 4$
23. $\frac{1}{3}x + \frac{1}{2}x = 5$
24. $\frac{3}{4}x - \frac{1}{2}x = 1$
25. $\frac{1}{5}x - \frac{1}{2}x = 3$
26. $3(2x-1) = 9x$
27. $5(x-1) = 2(1-x)$
28. $\frac{1}{5}(x+1) = 2$
29. $\frac{1}{3}(x+1) - 1 = 0$
30. $5 = \frac{1}{6}x - \frac{1}{3}x$
31. $8(1-3x) = 5(4x+6)$
32. $2 - x = x - 2$
33. $3 - 2(x-1) = 0$
34. $1 - \frac{1}{2}(1-x) = 0$
35. $\frac{1}{2}(x+5) - \frac{1}{4}(x-1) = 3$

3 Literal equations

An equation in a variable x, such as $ax + b = c + 1$ in which *letters* a, b, c are used to represent numerals, is called a *literal equation*. The members of the solution set will necessarily contain these letters.

Note: It will be assumed that in the process of solving a literal equation no step will imply division by zero. For example, in the fraction $\dfrac{a}{b-c}$, we assume that $b \neq c$.

Alternative forms of solution set are possible in some of the questions in this Section.

Algebra

Example 1. Solve for x, $ax+b = c$.   Compare

$$
\begin{aligned}
ax+b &= c & 2x+3 &= 4 \\
\Leftrightarrow \quad ax+b-b &= c-b & \Leftrightarrow \quad 2x+3-3 &= 4-3 \\
\Leftrightarrow \quad ax &= c-b & \Leftrightarrow \quad 2x &= 1 \\
\Leftrightarrow \quad \frac{1}{a} \times ax &= \frac{1}{a} \times (c-b) & \Leftrightarrow \quad \frac{1}{2} \times 2x &= \frac{1}{2} \times 1 \\
\Leftrightarrow \quad x &= \frac{c-b}{a} & \Leftrightarrow \quad x &= \frac{1}{2}
\end{aligned}
$$

The solution set is $\left\{\dfrac{c-b}{a}\right\}$.   The solution set is $\left\{\dfrac{1}{2}\right\}$.

Example 2. Solve for x, $a(x+b) = c$.   Compare

$$
\begin{aligned}
a(x+b) &= c & 3(x+2) &= 10 \\
\Leftrightarrow \quad ax+ab &= c & \Leftrightarrow \quad 3x+6 &= 10 \\
\Leftrightarrow \quad ax+ab-ab &= c-ab & \Leftrightarrow \quad 3x+6-6 &= 10-6 \\
\Leftrightarrow \quad ax &= c-ab & \Leftrightarrow \quad 3x &= 4 \\
\Leftrightarrow \quad \frac{1}{a} \times ax &= \frac{1}{a} \times (c-ab) & \Leftrightarrow \quad \frac{1}{3} \times 3x &= \frac{1}{3} \times 4 \\
\Leftrightarrow \quad x &= \frac{c-ab}{a} & \Leftrightarrow \quad x &= \frac{4}{3}
\end{aligned}
$$

The solution set is $\left\{\dfrac{c-ab}{a}\right\}$.   The solution set is $\left\{\dfrac{4}{3}\right\}$.

Example 3. Solve for x, $\dfrac{px}{q} = r$.   Compare

$$
\begin{aligned}
\frac{px}{q} &= r & \frac{5x}{6} &= 7 \\
\Leftrightarrow \quad \frac{q}{p} \times \frac{px}{q} &= \frac{q}{p} \times r & \Leftrightarrow \quad \frac{6}{5} \times \frac{5x}{6} &= \frac{6}{5} \times 7 \\
\Leftrightarrow \quad x &= \frac{qr}{p} & \Leftrightarrow \quad x &= \frac{42}{5}
\end{aligned}
$$

The solution set is $\left\{\dfrac{qr}{p}\right\}$.   The solution set is $\left\{\dfrac{42}{5}\right\}$.

Exercise 3

Solve for x, carefully setting out each step:

1 $x+4 = 7$ *2* $x+a = 7$ *3* $x+3 = b$ *4* $x+a = b$

Literal equations

5   $x-3=8$ 6 $x-a=5$ 7 $x-4=c$ 8 $x-p=q$

9   $x+h=4h$ 10 $x-c=5c$ 11 $x+2a=3a$

12   $x-2a=3a$ 13 $2x=5$ 14 $3x=a$

15   $ax=4$ 16 $cx=d$ 17 $px=-q$ 18 $-dx=f$

19   $kx+h=0$ 20 $2x+3=4$ 21 $2x+3a=4a$

22   $ax+5=9$ 23 $2x+a=b$ 24 $4x+r=h$

25   $3x-4=1$ 26 $3x-4a=a$ 27 $2x-c=d$

28   $ax-3=1$ 29 $ax-b=7$ 30 $px-q=r$

31   $2(x+3)=7$ 32 $3(x-b)=2b$ 33 $4(x+c)=d$

34   $5(x-h)=k$ 35 $a(x+b)=c$ 36 $p(x-q)=r$

37   $\dfrac{x}{2}=3$ 38 $\dfrac{x}{5}=a$ 39 $\dfrac{x}{c}=4$ 40 $\dfrac{x}{a}=b$

41   $\dfrac{x}{2}=\dfrac{3}{4}$ 42 $\dfrac{x}{a}=\dfrac{b}{2}$ 43 $\dfrac{2x}{3}=4$ 44 $\dfrac{3x}{4}=h$

45   $\dfrac{ax}{5}=b$ 46 $\dfrac{cx}{p}=q$ 47 $\dfrac{x}{a}=\dfrac{b}{c}$ 48 $\dfrac{1}{k}=\dfrac{x}{r}$

49   $\dfrac{x+a}{2}=5$ 50 $\dfrac{x+a}{b}=c$ 51 $\dfrac{x-p}{3}=q$ 52 $\dfrac{x-p}{q}=r$

Example. Solve for x, $a(x+b)=cx$

$$a(x+b)=cx$$
$$\Leftrightarrow \quad ax+ab=cx$$
$$\Leftrightarrow \quad ax-cx=-ab$$
$$\Leftrightarrow \quad x(a-c)=-ab$$
$$\Leftrightarrow \quad x=-\dfrac{ab}{a-c}$$

Note. Multiplying $\dfrac{-ab}{a-c}$ by $\dfrac{-1}{-1}$, i.e. 1, gives $\dfrac{ab}{-a+c}$, or $\dfrac{ab}{c-a}$.

This is an equivalent form of the solution.

Exercise 3B

Solve for x:

1   $x+a=4$ 2 $a-x=3$ 3 $2x+a=1$

Algebra

4. $a - 4x = 5$
5. $2x - a = b$
6. $5x + a = 3x + b$

7. $3x - a = x + b$
8. $(a+b)x = 3$
9. $a = bx + c$

10. $p(x+q) = r$
11. $a(x-c) = b$
12. $y = mx + c$

13. $y - x = 3$
14. $y + 2x = 3$
15. $5y - 2x + 10 = 0$

16. $4y + x + 5 = 0$
17. $2x + 3y + 4 = 0$
18. $\frac{x}{a} + \frac{y}{a} = 1$

19. $ax + by = c$
20. $\frac{x}{a} + \frac{y}{b} = 1$
21. $\tfrac{1}{3}(x+a) = 0$

22. $\tfrac{1}{2}(b-x) = a$
23. $ax + bx = 5$
24. $px - qx = 0$

25. $ax + b = cx + d$
26. $a(x-b) = cx$
27. $a(x-h) = b(x+k)$

28. $\dfrac{x-a}{x} = n$
29. $\dfrac{x+p}{x} = q$
30. $a = \dfrac{bx+b}{x}$

4 Cross-multiplication

An equation in the form of *two equal ratios*, such as $\dfrac{x}{a} = \dfrac{b}{c}$, occurs frequently in mathematics. Although the one shown may be solved by multiplying both sides by a, the following two-stage solution is worth studying.

$$\frac{x}{a} = \frac{b}{c}$$

$\Leftrightarrow ac \times \dfrac{x}{a} = ac \times \dfrac{b}{c}$

$\Leftrightarrow \quad cx = ab \quad \ldots (1)$

$\Leftrightarrow \dfrac{1}{c} \times cx = \dfrac{1}{c} \times ab$

$\Leftrightarrow \quad x = \dfrac{ab}{c} \quad \ldots (2)$

With practice the step marked (1) can be written down immediately. This process is called cross-multiplication, and is illustrated by the scheme:

$\dfrac{x}{a} \times\!\!\!\!\times \dfrac{b}{c} \iff cx = ab$

Cross-multiplication

Example 1

$$\frac{2x}{5} = \frac{3}{4}$$
$$\Leftrightarrow 8x = 15 \text{ (cross-multiplying)}$$
$$\Leftrightarrow x = \frac{15}{8}$$

Example 2

$$\frac{p}{x} = \frac{q}{r}$$
$$\Leftrightarrow qx = pr \text{ (cross-multiplying)}$$
$$\Leftrightarrow x = \frac{pr}{q}$$

Exercise 4

Solve the following equations by cross-multiplication, or otherwise:

1 $\dfrac{x}{4} = \dfrac{3}{1}$ *2* $\dfrac{x}{6} = \dfrac{1}{3}$ *3* $\dfrac{x}{4} = \dfrac{1}{2}$ *4* $\dfrac{x}{4} = \dfrac{3}{2}$

5 $\dfrac{x}{3} = \dfrac{3}{2}$ *6* $\dfrac{x}{5} = \dfrac{1}{2}$ *7* $\dfrac{3}{x} = \dfrac{1}{6}$ *8* $\dfrac{4}{5} = \dfrac{3}{x}$

9 $\dfrac{5x}{4} = 10$ *10* $\dfrac{2x}{8} = \dfrac{3}{4}$ *11* $\dfrac{5}{4x} = \dfrac{1}{3}$ *12* $\dfrac{0 \cdot 2x}{3} = \dfrac{3}{5}$

13 $\dfrac{x-1}{3} = \dfrac{5}{1}$ *14* $\dfrac{x+2}{2} = \dfrac{3}{1}$ *15* $\dfrac{2x-1}{2} = 1$ *16* $\dfrac{2x+9}{2} = 3$

17 $\dfrac{4-x}{3} = \dfrac{3}{2}$ *18* $\dfrac{2x+1}{2} = \dfrac{3}{5}$ *19* $\dfrac{2x-3}{4} = -\dfrac{1}{5}$ *20* $\dfrac{2-3x}{3} = -\dfrac{5}{6}$

Exercise 4B

Solve the following equations by cross-multiplication, or otherwise:

1 $\dfrac{x}{5 \cdot 7} = \dfrac{4}{3}$ *2* $\dfrac{3}{x} = \dfrac{5}{2 \cdot 4}$ *3* $\dfrac{x}{8} = \dfrac{2 \cdot 5}{2}$ *4* $\dfrac{2x}{5} = \dfrac{5 \cdot 4}{1 \cdot 8}$

5 $\dfrac{x}{a} = \dfrac{1}{2}$ *6* $\dfrac{x}{3} = \dfrac{b}{a}$ *7* $\dfrac{a}{x} = \dfrac{1}{a}$ *8* $\dfrac{b}{x} = \dfrac{c}{d}$

9 $\dfrac{1-x}{x} = \dfrac{a}{1}$ *10* $\dfrac{a-b}{x} = c$ *11* $\dfrac{a+x}{a-x} = \dfrac{3}{2}$ *12* $\dfrac{px-q}{px+q} = \dfrac{1}{2}$

13 $\dfrac{x-a}{x+a} = \dfrac{3}{4}$ *14* $\dfrac{ax+b}{ax-b} = \dfrac{3}{2}$ *15* $\dfrac{x+p}{x} = \dfrac{1}{q}$ *16* $\dfrac{x}{a-x} = \dfrac{m}{n}$

5 Changing the subject of a formula

Example 1. The formula $D = ST$ gives the distance D km covered by an object moving at an average speed of S km/h for T hours. Change the subject of the formula to S. Hence calculate the average speed of: *a* a car which travels 150 km in $2\frac{1}{2}$ hours *b* a space probe which travels the 380 000 km between the earth and the moon in 3 days.

$D = ST$

$\Leftrightarrow ST = D$

$\Leftrightarrow S = \dfrac{D}{T}$

a $S = \dfrac{D}{T}$

$= \dfrac{150}{2\frac{1}{2}}$

$= 60$

b $S = \dfrac{D}{T}$

$= \dfrac{380\,000}{3 \times 24}$

$= 5280$

The average speed of the car is 60 km/h, and of the space probe 5280 km/h.

Example 2. Change the subject of the formula $A = \pi r^2$ to r.

$A = \pi r^2$

$\Leftrightarrow \pi r^2 = A$

$\Leftrightarrow r^2 = \dfrac{A}{\pi}$

$\Leftrightarrow r = \sqrt{\dfrac{A}{\pi}}$

Example 3. A temperature in °F can be changed to °C by the formula $C = \frac{5}{9}(F-32)$. Make F the subject of this formula. Hence calculate F when: *a* $C = 0$ *b* $C = 100$.

$C = \frac{5}{9}(F-32)$

$\Leftrightarrow 9C = 5(F-32)$

$\Leftrightarrow 9C = 5F - 160$

$\Leftrightarrow 5F = 9C + 160$

$\Leftrightarrow F = \frac{1}{5}(9C+160)$

a $F = \frac{1}{5}(9C+160)$

$= \frac{1}{5}(0+160)$

$= 32$

b $F = \frac{1}{5}(9C+160)$

$= \frac{1}{5}(900+160)$

$= \frac{1}{5} \times 1060$

$= 212$

Changing the subject of a formula

Exercise 5

1. The perimeter of a square is $P = 4x$. Change the subject to x.

2. The area of a rectangle is $A = lb$. Change the subject to l.

3. The volume of a cuboid is $V = lbh$. Change the subject to h.

4. The speed of a train is $S = \dfrac{D}{T}$. Change the subject to: a D b T

5. The current in a circuit is $I = \dfrac{V}{R}$. Change the subject to: a V b R

6. The area of a triangle is $A = \tfrac{1}{2}bh$. Change the subject to h.

7. The area of a square is $A = x^2$. Change the subject to x.

8. The area of a circle is approximately $A = 3 \cdot 14 r^2$. Change the subject to r.

9. The area of a metal plate is $A = \tfrac{3}{2}ab$. Change the subject to a.

10. The equation of a straight line is $y = mx + c$. Change the subject to m.

11. The illumination from a lamp is $I = \dfrac{C}{d^2}$. Change the subject to:
       a C b d

12. The distance round a running track is $D = 2r(\pi + 1)$. Change the subject to r.

13. A number in a sequence is given by the formula $n = 2a + 3d$.
    a Change the subject of the formula to d.
    b Hence calculate d when $n = 30$ and $a = 6$.

14. The perimeter of a rectangle is $P = 2(l + b)$.
    a Change the subject of the formula to b.
    b Calculate b when $P = 22$ and $l = 7$.

15. Pythagoras' theorem for a right-angled triangle ABC can be stated as $a^2 = b^2 + c^2$.
    a Change the subject of this formula to c.
    b Calculate c when $a = 10$ and $b = 8$.

Algebra

16 The sum of the numbers in a series can be given by $S = \tfrac{1}{2}n(a+l)$.
  a Change the subject of this formula to l.
  b Calculate l when $S = 75$, $n = 20$ and $a = 5$.

17 The sum of the angles of a polygon with n sides is R right angles, where $R = 2n-4$.

  a Change the subject to n, and find how many sides a polygon has if its angle-sum is 10 right angles.
  b Can a polygon have an angle-sum of 15 right angles?

18 The volume of a cylinder, V cm^3, is given by $V = \pi r^2 h$, where r is the radius in cm, and h is the height in cm.

  a Make h the subject of the formula.
  b Make r the subject. Calculate r when $V = 33$, $h = 14$ and $\pi = \tfrac{22}{7}$.

Example. A formula for an electric current is $I = \dfrac{nE}{R+nr}$. Make n the subject of this formula.

$$I = \frac{nE}{R+nr}$$

$$\Leftrightarrow \quad I(R+nr) = nE$$
$$\Leftrightarrow \quad IR + Inr = nE$$
$$\Leftrightarrow \quad Inr - nE = -IR$$
$$\Leftrightarrow \quad n(Ir - E) = -IR$$
$$\Leftrightarrow \quad n = \frac{-IR}{Ir - E}, \text{ or } \frac{IR}{E - Ir}$$

Exercise 5B

1 The following formulae are used in dynamics. Change the subject in each case to the variable named.

  a $v = u + ft$; f
  b $s = ut + \tfrac{1}{2}ft^2$; u
  c $s = \tfrac{1}{2}(u+v)t$; t
  d $v^2 = u^2 + 2fs$; s
  e $s = ut + \tfrac{1}{2}ft^2$; f
  f $v^2 = u^2 + 2fs$; u

2 Make m the subject of the formula $\dfrac{m}{a} = \dfrac{b}{m}$.

Changing the subject of a formula

3. A formula for the length of an arc is $L = \frac{1}{3}(8h-c)$. Make h the subject.

4. $T = a+(n-1)d$ is a formula for the term T of a sequence of numbers. Change the subject of the formula to: *a* d *b* n

5. The volume of a sphere is $V = \frac{4}{3}\pi r^3$. Change the subject to r.

6. The area of a cylinder is $A = 2\pi r(r+h)$. Make h the subject.

7. Change the subject of the formula $y = \dfrac{1+x}{1-x}$ to x.

8. The profit $p\%$ on a sale is given by $p = \dfrac{100(s-c)}{c}$, where s and c are the selling price and cost price respectively.

   a Express s in terms of p and c.   *b* Express c in terms of p and s.

9. The sum of a series like $4+7+10+13+\ldots+22$ is given by the formula $S = \frac{1}{2}n(a+l)$, where S is the sum, n the number of terms that are added, a is the first term and l is the last term.

   a Verify that the formula is correct for the series given.
   b Find S when $a = 12, l = 72, n = 31$.
   c Make l the subject of the formula. Hence find the last term in a series of this kind with 20 terms whose first term is 8 and whose sum is 250.

10. The sum of a series like $32+16+8+4+\ldots$ to an 'infinite' number of terms is given by the formula $S = \dfrac{a}{1-r}$, where S is the sum, a the first term, and r the ratio of any term to the preceding term.

    a Use the formula to find the sum of the given series.
    b By adding the terms find the sum of 4 terms; also of 5 terms, 6, 7, 8, 9, and 10 terms.
    c Make r the subject of the formula. Hence calculate r when $S = 121\frac{1}{2}$ and $a = 81$.

Algebra

Summary

1. A *formula* is an equation which expresses one symbol (called the *subject of the formula*) in terms of one or more symbols;

   e.g. $A = bh$

2. The *subject of a formula* may be changed to put the formula in a form more suitable for the calculation required;

   e.g. $A = bh \Leftrightarrow h = \dfrac{A}{b}$

3. A *literal equation* is an equation in which letters are used to represent numerals and variables;

   e.g. $y = mx + c$

   x, y are variables; m, c are numerals.

Revision Exercises

Revision Exercises on Chapter 1
Further Sets and Graphs

Revision Exercise 1A

1. State whether each of the following is a well-defined set. If it is, say whether it is finite, infinite, or the empty set.
    - a The set of members of the staff of your school.
    - b The set of even positive integers.
    - c The set of geometrical figures which are symmetrical about a line.
    - d The set of prime numbers between 48 and 52.
    - e The set of fair-haired children in your town.

2. List the following sets, using dots when necessary:
    - a The set of the first fifty natural numbers.
    - b The set of vowels in the word *inequation*.
    - c The set of negative integers greater than -25.
    - d The set of all numbers of the form $3n$, where $n \in N$.

3. $E = \{A, E, F, H, I, L, M, N, O, T, V, W, X, Y, Z\}$
    - a List the set of capital letters in E which are not altered when they are turned upside down.
    - b List the set of letters which are not altered when they are turned over sideways.
    - c List the intersection of these two sets.

4. $E = \{\text{all triangles}\}$; $A = \{\text{isosceles triangles}\}$, $B = \{\text{equilateral triangles}\}$, $C = \{\text{right-angled triangles}\}$, $D = \{\text{triangles all of whose angles are equal}\}$.
    - a Which of the following statements are true, and which are false?
      $$B \subset A, \ A = C, \ B = D, \ A \subset D$$
    - b (*1*) If $c \in A$ and $c \in C$, give the size of the angles of c.
      (*2*) If $f \in D$, what can you say about the sides of f?

Algebra

5 $E = \{1, 2, 3, \ldots, 20\}$ and x is a variable on E.
- **a** Write in list form $\{x: x \text{ is even}\}$.
- **b** List the members of the subset of E such that each member is exactly divisible by 4.
- **c** List the elements of $\{x: x \text{ is prime}\}$.

6 $L = \{a, b, c, d, e\}$. Write down:
- **a** Three subsets of L containing three members.
- **b** Four subsets of L containing four members.
- **c** A subset of L which is a subset of all seven of these sets.

7 $E = \{1, 2, 3, \ldots, 10\}$; $H = \{1, 2, 3, 4\}$ and $K = \{4, 5, 6, 7, 8\}$.
- **a** Describe, by listing the elements, the sets $H \cap K$ and $H \cup K$.
- **b** Illustrate by a Venn diagram.
- **c** Show that $n(H \cup K) + n(H \cap K) = n(H) + n(K)$.

8 Simplify each of the following:
- **a** $\{1, 2, 3, 4\} \cap \{3, 5, 7\}$
- **b** $\{-1, 0, 1\} \cap \emptyset$
- **c** $\{1, 2, 3\} \cup \{1, 3, 4\}$
- **d** $\{3, 1, 2\} \cup \{2, 3, 1\}$

9 From the information given in Figure 1, write in list form each of the following sets:
- **a** A
- **b** B
- **c** A'
- **d** B'
- **e** $A \cap B$
- **f** $A \cup B$
- **g** $A' \cap B'$
- **h** $A' \cup B'$
- **i** $(A \cup B)'$
- **j** $(A \cap B)'$

Which of these are equal sets?

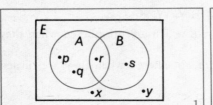
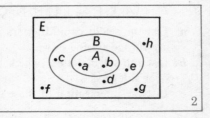

10 Repeat question **9** for the sets shown in Figure 2.

11 Draw the graph of each of the following mappings from R to R. A suggested replacement set for x is given to help you to draw the complete graph.

- **a** $x \to x-2$, $x \in \{-1, 0, 1, 2, 3, 4\}$.
  Scale: 1 cm to 1 unit on each axis.

Revision Exercises on Chapter 1

  b $x \to 2-x$, $x \in \{-2, -1, 0, 1, 2, 3\}$.
     Scale: 1 cm to 1 unit on each axis.
  c $x \to x^2+2$, $x \in \{-3, -2, -1, 0, 1, 2, 3\}$.
     Scale: 1 cm to 1 unit for x, 1 cm to 2 units for x^2+2.
  d $x \to 6-x^2$, $x \in \{-3, -2, -1, 0, 1, 2, 3\}$.
     Scale: 1 cm to 1 unit for x, 1 cm to 2 units for $6-x^2$.

Revision Exercise 1B

1   In this question a special notation is used. The natural numbers up to and including 32 are each in turn divided by 4. $\bar{2}$ is the set of such numbers which give a remainder 2 on division by 4, and similarly for $\bar{0}, \bar{1}, \bar{3}$.
List the sets $\bar{0}, \bar{1}, \bar{2}, \bar{3}$ using dots to signify 'and so on'. Complete the following:

  a If $a \in \bar{1}$ and $b \in \bar{2}$, $a+b \in \ldots$ b If $a \in \bar{0}$ and $b \in \bar{3}$, $a+b \in \ldots$
  c If $a \in \bar{2}$ and $b \in \bar{3}$, $a+b \in \ldots$ d If $a \in \bar{1}$ and $b \in \bar{3}$, $a+b \in \ldots$

2   Which of the following are equal sets and which are subsets of others in the question?

  $P = \{$letters of the word MAD$\}$
  $Q = \{$prime factors of 30$\}$ $R = \{$prime factors of 90$\}$
  $S = \{$letters of the word MADMAN$\}$
  $T = \{$letters of the word MADAM$\}$
  $V = \{$prime numbers less than 10$\}$

3 $E = \{1, 2, 3, \ldots, 12\}$, $P = \{1, 2, 3, 4, 5, 6, 7\}$, $Q = \{5, 6, 7, 8, 9, 10\}$.
List the members of the following sets:

  a $P \cap Q$ b $P \cup Q$ c P' d Q'
  e $(P \cap Q)'$ f $(P \cup Q)'$ g $P \cap Q'$ h $P' \cap Q$

Draw a Venn diagram showing the sets E, P, Q.

4   The subsets of $E = \{a, b\}$ are $A = \{a\}$, $B = \{b\}$, $E = \{a, b\}$ and ø.
Copy and complete the following tables of union and intersection.

$\cup$	ø	A	B	E		$\cap$	ø	A	B	E
ø	ø	A	.	.		ø	ø	.	.	.
A	.	.	E	.		A	.	.	.	.
B	.	.	.	.		B	.	.	B	.
E	.	.	.	.		E	.	A	.	.

Algebra

In what way do these tables of union and intersection of sets resemble the tables of addition and multiplication of whole numbers?

5   In Figure 3, $\mathscr{A}$ is the set of points on the perimeter of a square and $\mathscr{C}$ is the set of points on the circumference of a circle. If $\mathscr{P} = \mathscr{A} \cap \mathscr{C}$, we see that $\mathscr{P} = \{X, Y\}$ and $n(\mathscr{P}) = 2$. Draw diagrams to show the relative positions of the square and circle in each of the following, and list the set $\mathscr{P}$ (letter your diagram) in each case:

  a $n(\mathscr{P}) = 0$ *b* $n(\mathscr{P}) = 1$ *c* $n(\mathscr{P}) = 3$
  d $n(\mathscr{P}) = 4$ *e* $n(\mathscr{P}) = 8$ *f* $n(\mathscr{P}) = 6$

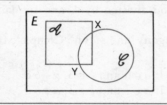
3
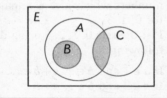
4

6 *a* Write down the set represented by the shaded area in Figure 4, using the symbols ∪ and ∩.
  b Draw a Venn diagram for three non-empty sets P, Q, R such that P, Q, R have the following properties: $R \subset Q$, $Q \neq R$ and $P \cap R = \emptyset$.

7   The universal set E is the set of integers, and x is a variable on E.

  a List the sets (*1*) $P = \{x : 0 \leqslant x \leqslant 4\}$
      (*2*) $Q = \{x : -2 \leqslant x+1 \leqslant 3\}$ (*3*) $R = \{x : -5 \leqslant 3x+1 \leqslant 10\}$
  b If $P \cup Q \cup R = \{x : a \leqslant x \leqslant b\}$, find a and b.
  c Show that $P \cap (Q \cup R) = (P \cap Q) \cup (P \cap R)$

8   $P, Q, S, T, G, H, A,$ and B are sets. Answer the following by first drawing a Venn diagram for each and then entering the number of elements in the appropriate regions.

  a If $n(P) = 76$, $n(Q) = 24$ and $n(P \cap Q) = 10$, find $n(P \cup Q)$.
  b If $n(S) = 83$, $n(T) = 17$ and S and T are disjoint, find $n(S \cup T)$.
  c If $n(G) = 57$, $n(H) = 69$ and $G \subset H$, find $n(G \cup H)$.
  d If $n(A) = 86$, $n(B) = 70$ and $n(A \cup B) = 140$, find $n(A \cap B)$.

Revision Exercises on Chapter 1

9 A, B, C are three sets whose intersections are not empty sets. Draw a Venn diagram to illustrate this information. From the data in the following table, calculate $n(A \cup B \cup C)$.

Set	A	B	C	$A \cap B$	$A \cap C$	$B \cap C$	$A \cap B \cap C$
Number of elements	46	40	50	14	22	9	4

10 To cater for a school party, all of the 115 pupils involved brought at least one of these: sandwiches, cakes, or lemonade. 70 brought lemonade and 54 brought sandwiches. 19 brought lemonade and cakes only, and 22 brought lemonade and sandwiches only. 24 brought lemonade only, and 15 brought sandwiches only. How many brought cakes only?

11 The universal set E is the set of all points inside a given circle. E and its subsets A, B, C, D are shown by the shaded regions in Figure 5. Draw a diagram to show by shading each of the following sets:

a A' b C' c $A \cup D$ d $A \cap B$
e $A' \cap C'$ f $(A \cup D)'$ g $C \cap D$ h $(A \cap B) \cup D$

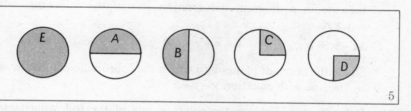

5

12 Each of the sets below is represented by one of the eight regions numbered I to VIII in Figure 6.

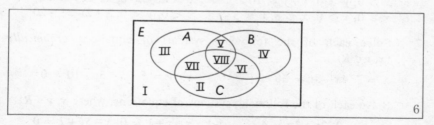

6

a $A \cap B \cap C$ b $A \cap B' \cap C$ c $A \cap B \cap C'$
d $A \cap B' \cap C'$ e $A' \cap B \cap C$ f $A' \cap B \cap C'$
g $A' \cap B' \cap C$ h $A' \cap B' \cap C'$

Match each of these sets to its region.

Algebra

Revision Exercises on Chapter 2
Systems of Equations and Inequations in Two Variables

Revision Exercise 2A

1. List the set of ordered pairs which make the open sentence $2x+y = 7$ a true sentence, where $x, y \in W$. Show this solution set in a diagram.

2. Repeat question *1* for the inequation $x+2y \leqslant 4$.

3. Rearrange each of the following equations in the form:
   (i) $ax+by+c = 0$ (ii) $y = mx+c$:
   a $2x+y = 3$ *b* $x+2y = -5$ *c* $2x-3y = 9$

4. Indicate in a diagram the solution sets of the following, where $x, y \in R$:
   a $2x-y = 8$ *b* $x+y \leqslant 10$ *c* $3x+4y = 24$

5. Shade the two half planes into which the coordinate plane is divided by the line with equation $x-y-5 = 0$.

6. Indicate in a diagram the solution sets of the following systems $(x, y \in R)$:
   a $x \geqslant 0, y \geqslant 0, x+y \leqslant 10$ *b* $x > 0, y > 0, 2x+5y < 20$
   c $x \geqslant 0, y \geqslant 0, x \leqslant 8, y \leqslant 3$ *d* $x < 0, y < 0, x+2y > -20$

7. Solve each of the following systems of equations *graphically* $(x, y \in R)$:
   a $x = 7, 4x-3y = 36$ *b* $x-y = 5, 2x+5y-10 = 0$

8. Solve each of the following systems of equations, where $x, y \in R$:
   a $x-y = 2, 2x+3y = 4$ *b* $x+y+1 = 0, x-5y+7 = 0$
   c $y = 2x+3, 3x+4y-1 = 0$ *d* $2x+5y = 14, 3x-2y = -17$

9. Solve each of the following systems of equations, where $x, y \in R$:
   a $x+2y = 8, \frac{3}{4}x+\frac{1}{2}y = 4$ *b* $3x = 4y, \frac{1}{2}x+\frac{1}{3}y = 1$

10 Figure 7 shows the line with equation $2x+y = 8$.
  a Give the coordinates of the points A and B.
  b Which regions are given by $\{(x, y): 2x+y \geq 8\}$?
  c Use set notation to define the region AOB.

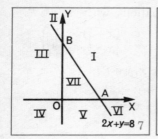
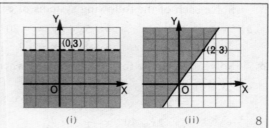

11 For each diagram in Figure 8 write down an inequation whose solution set is indicated by the shaded region.

12 $x^2 + ax + b = 0$ when $x = 2$ and also when $x = -1$. Obtain two equations in a and b, and solve them for a and b.

13 If $y = px + q$ and $y = -3$ when $x = 1$, and $y = 9$ when $x = 3$, find p and q.

14 $S = \{(x, y): 2x + 3y = 0\}$, $T = \{(x, y): 4x - y = 7\}$, $x\ y \in R$. Find $S \cap T$.

Revision Exercise 2B

1 Sketch the solution sets of the following, where $x, y \in R$:
  a $0 \leq x \leq 5$ b $x+y < 12$ c $y \leq 3x - 15$ d $-2 \leq y \leq 2$

2 Indicate in a diagram the solution sets of the following systems, where $x, y \in R$:
  a $y \leq 8, y \geq x, 2x+y \geq 4$ b $x > 0, y < 0, 3x - 2y < 18$
  c $x \geq 0, y \geq 0, 2x+y \leq 10, x+2y \leq 10$
  d $x \leq 8, y \geq 5, y \leq x+5$

3 Rearrange each of the following equations in the form
  (i) $ax + by + c = 0$ (ii) $y = mx + c$:
  a $5x - 2y = 6$ b $5 - 3x = 7y$ c $\tfrac{1}{2}x + \tfrac{1}{3}y = \tfrac{1}{4}$

Algebra

4 Solve each of the following systems of equations graphically $(x, y \in R)$:
- a $\quad x+y = 4,\ 2x-y+1 = 0$
- b $\quad 2x+3y = 6,\ 2x+3y = -6$
- c $\quad x-y+2 = 0,\ 2x+y-2 = 0,\ 3x-y+2 = 0$

5 Solve each of the following systems of equations, where $x, y \in R$:
- a $\quad \begin{aligned} x+y &= 0 \\ 2x+3y &= 6 \end{aligned}$
- b $\quad \begin{aligned} 5x+6y &= -20 \\ 3x-4y &= 26 \end{aligned}$
- c $\quad \begin{aligned} \tfrac{1}{4}x+\tfrac{1}{3}y &= \tfrac{5}{4} \\ \tfrac{2}{3}x-\tfrac{1}{5}y &= \tfrac{1}{15} \end{aligned}$

6 Solve each of the following systems, where $x, y \in R$:
- a $\quad \begin{aligned} \tfrac{1}{4}x+\tfrac{1}{6}y &= 1 \\ 4x-y &= 5 \end{aligned}$
- b $\quad \begin{aligned} \tfrac{1}{4}(x-1)+y &= 8 \\ \tfrac{1}{6}(y-1)+x &= 6 \end{aligned}$
- c $\quad \dfrac{x}{2}+\dfrac{y}{2} = 1,\ \dfrac{2x+4y}{5}-\dfrac{x-y}{3} = -2$

7 Figure 9 shows the lines with equations $x = 5$, $y = 3$ and $y = x+3$.
- a Write down the coordinates of P, Q and R.
- b Give a set definition of the region PQR.
- c Give a set definition of the half plane bounded by the line of PR, which does not contain the origin.

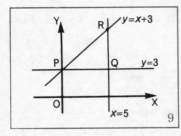

9

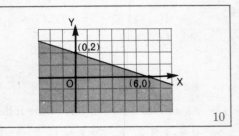

10

8 Write down an inequation whose solution set is indicated by the shaded region in Figure 10.

9 Points $(1, 2)$ and $(-1, -4)$ lie on the straight line with equation $y = mx+c$. By substituting the coordinates of the points in this equation obtain two equations in m and c. Solve these equations for m and c.

10 The formula for finding the area of a circular ring is
$$A = \tfrac{22}{7}(R+r)(R-r).$$
If $A = 44$ and $R+r = 7$, find r.

Revision Exercises on Chapter 3

11 Marks which range from 24 to 88 are to be scaled so as to range from 20 to 100. An original mark x becomes a scaled mark y where $y = px+q$, and p and q are constants. Find p and q.
  a What is the scaled mark for an original mark of 60?
  b What original mark gives the same scaled mark?

12 If $5x-y = 8$ and $5y-x = 20$, find the value of x^2+y^2.

13 If $a = 3b$ and $b+c = 1$, find ab when $c = 5$.

14 For the series $8+12+16+20+...$ we can use the notation $S_1 = 8$, $S_2 = 8+12$, $S_3 = 8+12+16$, etc., to denote the sum of terms. If $S_n = an^2+bn$, put $n = 1$ and $n = 2$ to obtain equations which you can use to find a and b.
  Hence find the sum of $8+12+16+20+...$ to 100 terms.

Revision Exercises on Chapter 3
Formulae and Literal Equations

Revision Exercise 3A

1 Figure 11 shows a cuboid x cm long, y cm wide and y cm high. Write down a formula for: a the total length L cm of the edges b the total surface area A cm² c the volume V cm³ of the cuboid.
  Given that $x = 8 \cdot 5$ and $y = 5$, calculate L, A and V.

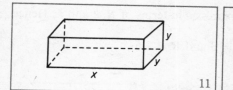
11

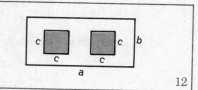
12

2 Figure 12 shows a rectangular steel plate of length a cm and breadth b cm from which two squares of side c cm are removed. Find a formula for the area A cm² of the remainder of the plate. Calculate A when $a = 50$, $b = 40$ and $c = 25$.

Algebra

3 Solve for x:

a $x+p = q$ b $x-p = 3p$ c $px = 1$ d $\tfrac{1}{2}x = n$

e $\dfrac{x}{c} = t$ f $2x = a-b$ g $a-x = b$ h $p(x-q) = r$

4 a Write down the sums of the following series of numbers:
   $1+3$, $1+3+5$, $1+3+5+7$, $1+3+5+7+9$.
   b From the pattern of your answers, write down a formula for the sum N of the first n odd numbers. Check the accuracy of your formula by working out the sum of the first ten odd numbers.

5 In each of the following equations, express y in terms of x:

a $y-2x+3 = 0$ b $4x+2y = 3$ c $3x-2y = 6$

d $\tfrac{1}{2}x+y = 4$ e $\dfrac{x}{y} = \dfrac{4}{5}$ f $\dfrac{x}{3}+\dfrac{y}{2} = 1$

6 A car costing £m is sold to a customer for £n. The firm makes a profit of £$\tfrac{1}{4}m$. Show that $n = \tfrac{5}{4}m$.
Change the subject of this formula to m, and hence find m when $n = 1200$.

7 Make r the subject in each of the following formulae:

a $C = 2\pi r$ b $A = \pi r^2$ c $V = \pi r^2 h$ d $a = \dfrac{v^2}{r}$ e $V = \tfrac{4}{3}\pi r^3$

8 If $x = 3a+b$, $y = a-3b$ and $P = 4x-3y$, express P in terms of a and b. Hence find P given that $a = 3\cdot 4$ and $b = 0\cdot 8$.

9 The formula $Ft = m(v-u)$ occurs in mechanics. Change the subject to: a m b v c u.

10 From the formula $R = \dfrac{kL}{d^2}$, express d in terms of R, k and L. Hence calculate d when $k = 7\cdot 5$, $L = 3$ and $R = 2\cdot 5$.

Revision Exercise 3B

1 Solve for x:

a $p+qx = r$ b $x-a = b-x$ c $mx+nx = p$

d $ax+b = c+x$ e $\dfrac{x}{a}-b = \dfrac{x}{c}$ f $\dfrac{x}{a}-a = \dfrac{x}{b}-b$

g $ax+b(x-a) = ab$ h $a(x+b) = b(x+a)$ i $m(x-n) = n(x+m)$

Revision Exercises on Chapter 3

2 Find a formula for the nth term of each of these sequences. For example, the nth term of the sequence $1^2, 2^2, 3^2, 4^2, \ldots$ is n^2.

  a $1, 2, 3, 4, \ldots$ *b* $3, 9, 27, 81, \ldots$

  c $\dfrac{1}{1\times 2}, \dfrac{1}{2\times 3}, \dfrac{1}{3\times 4}, \dfrac{1}{4\times 5}, \ldots$

  Use your formulae to give the 10th term of each sequence.

3 A man's income of £m is subject to income tax at the rate of c pence per £ of income. Find a formula for the balance £p after income tax has been deducted. Express your formula in factorised form.

4 A cylindrical rod of length L metres is divided into two parts.

  a If the length of one of these parts is x metres, write down the length of the other part in terms of L and x. Show these in a sketch.

  b Given that the ratio of the two parts is $m:n$, find a formula for x in terms of L, m and n.

5 In each of the following, express y in terms of x:

  a $2x+y = 3$ *b* $2x-y = 5$ *c* $x-3y+9 = 0$

  d $2x+3y+12 = 0$ *e* $\dfrac{x}{3}+\dfrac{y}{4} = 1$ *f* $\dfrac{y+1}{x-3} = \dfrac{4}{3}$

6 In Figure 13, OQ bisects angle POR. If angle POQ $= x°$ and angle ROS $= y°$, find a formula for:

  a y in terms of x *b* x in terms of y.

  If $x < 50$, obtain the simplest inequation in y other than $y > 0$.

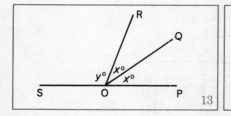

13

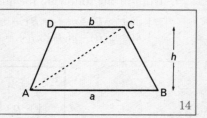

14

7 In Figure 14, the parallel sides of trapezium ABCD are a cm and b cm long and the distance between them is h cm.

Algebra

    a Write down expressions for the area of triangles ABC and ACD, and hence show that the area A cm² of trapezium ABCD is given by formula $A = \frac{1}{2}(a+b)h$.
    b Change the subject of this formula to h.
    c Calculate h when $A = 40\cdot5$, $a = 12\cdot4$ and $b = 5\cdot6$.

8   Transform each of the following formulae to the symbols named:
    a $Q = \frac{1}{2}(10-x)$ to x *b* $v = u - gt$ to t
    c $E = \dfrac{9nk}{3k+n}$ to n *d* $s = \dfrac{W}{W-w}$ to (*1*) w and (*2*) W

9   The volume V cm³ of a cone is given by the formula $V = \frac{1}{3}\pi r^2 h$, where r cm is the radius of the base and h cm is the height.
    a Make h the subject of the formula.
    b Make r the subject. Calculate r when $V = 157$, $h = 5$ and $\pi = 3\cdot14$.

10   The square of the number $n+\frac{1}{2}$ can be calculated using the formula $(n+\frac{1}{2})^2 = n(n+1) + \frac{1}{4}$. Use this formula to calculate mentally:
    a $(4\frac{1}{2})^2$ *b* $(11\frac{1}{2})^2$ *c* $(3\cdot5)^2$ *d* $(9\cdot5)^2$ *e* 75^2

11*a* If $\dfrac{1}{v} - \dfrac{1}{u} = \dfrac{1}{20}$, express v in terms of u by first multiplying both sides of the equation by $20uv$.
    b Make v the subject of the formula $\dfrac{1}{u} + \dfrac{1}{v} = \dfrac{2}{r}$.

12   The Venn diagram in Figure 15 shows the relation between sets A, B and E. The number of elements in A is $(p+q)$, in B it is $(q+r)$, and the number of elements that do not belong to A or B is s. Check the truth of the following formulae:
    (*1*) $n(A \cap B) + n(A \cup B) = n(A) + n(B)$
    (*2*) $n(A \cup B) + n(A' \cap B') = n(E)$

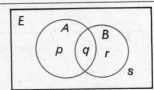

15

Cumulative Revision Section (Books 1-4)

Book 1 Chapter Summaries

Chapter 1 An introduction to sets

1. A *set* is a collection of clearly defined objects. $A = \{a, b, c\}$
2. Each object in a set is a *member* or *element* of that set, and *belongs* to the set. $a \in A, p \notin A$
3. *Equal sets* have exactly the same members. $B = \{c, a, b\}, A = B$
4. The *empty set* is the set with no members. $\emptyset$ or $\{\ \}$
5. A *universal set* is the set of all elements being discussed. E('Entirety')
6. A set B is a *subset* of a set A if every element of B is a member of A. $B \subset A, A \subset A$
   $\emptyset \subset A, \emptyset \subset E$
7. The *intersection* of two sets A and B is the set of elements which are members of A and are also members of B. $A \cap B$
8. *Venn diagrams*:

$A \cap B$

$A \subset B$

$A \cap B = \emptyset$

$A = B$

Chapter 2 Mathematical sentences

1. A *sentence* may be true or false.
2. An *open sentence* is a sentence containing a variable or variables, e.g. '12 is divisible by x'; '$x + y = 5$'.
3. A *variable* is a symbol which can be replaced by members of a given set.

Algebra

4   The *solution set* of an open sentence is the set of replacements of the variable which give a true sentence.

       Each member of the solution set is a *solution* of the open sentence.

5   An *equation* is an open sentence containing the verb 'is equal to'.

6   Mathematical sentences can be represented by *graphs*.

Chapter 3 Multiplication

1   *Multiplication is commutative*
      e.g. $3 \times a = a \times 3$, and is written $3a$.

2   *Multiplication is associative*
      e.g. $(3 \times a) \times b = 3 \times (a \times b) = 3ab$.

3   *Multiplication is distributive over addition*
      e.g. $a(b+c) = ab+ac$.

4   *The coefficient in a term is the numerical factor in the term*
      e.g. the coefficient of $3a$ is 3.

5   *The value* of $3x^2$ when $x = 2$ is $3 \times 2 \times 2 = 12$.

6    $a \times a = a^2$, and $a \times a \times a = a^3$.
      Also $a+a = 2a$, $a+a+a = 3a$.

Book 2 Chapter Summaries

Chapter 1 Replacements and formulae

1   *Replacements* can be made for the variable in an open sentence.

2   *Formulae* are useful when the same kind of calculation has to be repeated with different sets of numbers.

     Examples are:

     Perimeter of rectangle. $P = 2l+2b$, or $P = 2(l+b)$
     Area of rectangle. $A = lb$
     Volume of cuboid. $V = lbh$

Book 2 Chapter Summaries

Chapter 2 Inequalities and inequations

1. The numbers used for counting and measuring have *order*, which can be shown on the number line.

2. If a and b are any two numbers, then:
$$a > b, \quad \text{or} \quad a = b, \quad \text{or} \quad a < b$$

3. | Inequality symbols | Meaning |
   |---|---|
   | $>$ | is greater than |
   | $\geqslant$ | is greater than or equal to |
   | $<$ | is less than |
   | $\leqslant$ | is less than or equal to |
   | $\neq$ | is not equal to |

4. An *inequation* is an open sentence containing one of the verbs in *3* above.

                                    Inequation
   5 > 3 (True) $x > 3$ (neither true nor false)
   5 < 3 (False)

5. A *solution* of an inequation is a replacement for the variable which gives a true sentence, e.g. 2 is a solution of $x < 3$.
   The *solution set* is the set of all such solutions in a given set E.

6. The *graph* of an inequation in one variable shows its solution set on the number line, e.g. for $x < 3$, where $x \in W$, we have:

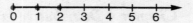

Chapter 3 Negative numbers

1. a. *Positive integers*: $+1, +2, +3, \ldots$; or $1, 2, 3, \ldots$
      A positive integer a is *greater* than zero; $a > 0$
   b. *Negative integers*: $\ldots, -3, -2, -1$.
      A negative integer b is *less* than zero; $b < 0$
   c. *Zero* is neither positive nor negative.

Algebra

    d The set of integers $Z = \{..., -3, -2, -1, 0, 1, 2, 3, ...\}$
    e Q is the set of *rational numbers*
        $\frac{3}{4}$, 2·4, $-\frac{1}{2}$, $\frac{10}{5}$ are rational numbers.

2   *Addition* of numbers:
    a $5+3 = 8$ *b* $5+(-3) = 2$
    c $-5+3 = -2$ *d* $-5+(-3) = -8$

3   Addition of numbers is:
    (i) commutative $a+b = b+a$
    (ii) associative $(a+b)+c = a+(b+c)$

4   *Negatives* (or *additive inverses*)
    a and $(-a)$ are *negatives* of each other, and are such that
$$a+(-a) = (-a)+a = 0$$

5   *Subtraction*: To subtract b from a, *ADD* the negative of b to a.
            $a-b = a+(-b)$
e.g. $4-9 = 4+(-9) = -5$
    Thus, every subtraction may be thought of as an addition.

6   *Standard form* $a \times 10^n$, where $1 \leqslant a < 10$, and n is a positive or negative integer. Examples:
    a $342 = 3·42 \times 100 = 3·42 \times 10^2$
    b $0·045 = \dfrac{4·5}{100} = \dfrac{4·5}{10^2} = 4·5 \times 10^{-2}$

Chapter 4 Distributive law

1   *Products to sums or differences: multiplying out.*
    (i) $a(b+c) = ab+ac$
    (ii) $a(b-c) = ab-ac$
    (iii) $\dfrac{1}{a}(b+c) = \dfrac{b}{a}+\dfrac{c}{a}$

2   *Sums or differences to products: common factors.*
    (i) $ab+ac = a(b+c)$
    (ii) $ab-ac = a(b-c)$ a is a *common factor* of ab and ac.

> Summary

3   Collecting like terms.

In $4x+3y+2x+y$, $4x$ and $2x$ are like terms, and $3y$ and y are like terms.

$$4x+3y+2x+y$$
$$= 4x+2x+3y+y$$
$$= 6x+4y$$

Book 3 Chapter Summaries

Chapter 1 Relations, mappings and graphs

1   A *relation* from a set A to a set B is a pairing of elements of A with elements of B. If $B = A$, the relation is said to be *on A*.

2   A *relation* may be completely described by:
   (1) A *set of ordered pairs*.
   (2) The *solution set of an open sentence*, such as 'x owns y', which gives the ordered pairs in the relation.
   (3) An *arrow diagram* or a *Cartesian graph*.

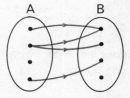

A relation which is not a mapping.

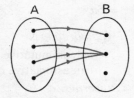

A relation which is a mapping.

3   A *mapping* from a set A to a set B is a relation in which every element of A is paired with exactly one element of B.
   The element of B is called the *image* of the element of A under the mapping.

4   Two sets A and B are said to be in *one-to-one correspondence* if the elements of A and B can be paired so that each element

Algebra

of A corresponds to exactly one element of B, and each element of B corresponds to exactly one element of A.

Example. $A = \{1, 2, 3, ..., 12\}$, $B = \{$months of the year$\}$

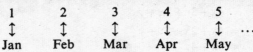

Chapter 2 Operations on integers and rational numbers

1 *Addition* is

    (i) a commutative operation: $a+b = b+a$
    (ii) an associative operation: $(a+b)+c = a+(b+c)$

$a+0$ $-3y+(-7y)$ $-\frac{3}{4}+\frac{1}{2}$
$= 0+a$ $= -10y$ $= -\frac{3}{4}+\frac{2}{4}$
$= a$ $= -\frac{1}{4}$

2 *Subtraction.* Subtracting b from a is the same as adding the negative of b to a, i.e. $a-b = a+(-b)$

$9a-4a$ $9a-(-4a)$ $-9a-(-4a)$
$= 9a+(-4a)$ $= 9a+4a$ $= -9a+4a$
$= 5a$ $= 13a$ $= -5a$

3 *Multiplication* is

    (i) a commutative operation: $ab = ba$
    (ii) an associative operation: $(ab)c = a(bc)$

Properties. $a \times 0 = 0 \times a = 0$; $a \times 1 = 1 \times a = a$

The product of two positive or two negative integers or rational numbers is positive; the product of a positive and negative number is negative.

$6 \times \frac{2}{3}$ $6 \times (-\frac{2}{3})$ $-6 \times \frac{2}{3}$ $-6 \times -\frac{2}{3}$
$= 4$ $= -4$ $= -4$ $= 4$

4 *Multiplication* is distributive over addition:

$a(b+c) = ab+ac$,   or $ab+ac = a(b+c)$

$2(x-3)$ $-3(x-2y)$ $5x-(2x+3)$
$= 2x-6$ $= -3x+6y$ $= 5x-2x-3$
                                               $= 3x-3$

Book 3 Chapter Summaries

5 *Division.* If a and b are non-zero rational numbers, dividing a by b is the same as multiplying a by the reciprocal (or multiplicative inverse) of b, i.e. $a \div b = a \times \dfrac{1}{b}$

Division by zero is not possible.

$$\begin{array}{llll} 6 \div \tfrac{2}{3} & 6 \div (-\tfrac{2}{3}) & -6 \div \tfrac{2}{3} & -6 \div (-\tfrac{2}{3}) \\ = 6 \times \tfrac{3}{2} & = 6 \times (-\tfrac{3}{2}) & = -6 \times \tfrac{3}{2} & = -6 \times (-\tfrac{3}{2}) \\ = 9 & = -9 & = -9 & = 9 \end{array}$$

Chapter 3 Equations and inequations in one variable

1 *Addition principle*

The same number may be added to each side of an equation or inequation to give an equivalent equation or inequation, i.e. one which has the same solution set.

$$\begin{array}{ll} x+3 = 5,\ x \in W & x-2 < -5,\ x \in Z \\ \Leftrightarrow x+3+(-3) = 5+(-3) & \Leftrightarrow x-2+2 < -5+2 \\ \Leftrightarrow \qquad x = 2 & \Leftrightarrow \qquad x < -3 \end{array}$$

2 *Multiplication principle*

a Each side of an *equation* may be multiplied by the same non-zero number to give an equivalent equation.

b Each side of an *inequation* may be multiplied by the same *positive* number to give an equivalent inequation; but if the multiplier is a *negative* number the inequality symbols must be reversed.

$$\begin{array}{ll} 5x = 3,\ x \in Q & 5x > 3,\ x \in Q \\ \Leftrightarrow \tfrac{1}{5} \times 5x = \tfrac{1}{5} \times 3 & \Leftrightarrow \tfrac{1}{5} \times 5x > \tfrac{1}{5} \times 3 \\ \Leftrightarrow \qquad x = \tfrac{3}{5} & \Leftrightarrow \qquad x > \tfrac{3}{5} \end{array}$$

$$\begin{array}{c} -5x > 3,\ x \in Q \\ \Leftrightarrow -\tfrac{1}{5} \times (-5x) < -\tfrac{1}{5} \times 3 \\ \Leftrightarrow \qquad x < -\tfrac{3}{5} \end{array}$$

Algebra

3 Solving equations and inequations

To solve an equation or inequation we have to find the simplest equivalent equation or inequation.

To do this we arrange that variables are on one side (usually the left side) and constants are on the other side, by *adding an appropriate negative to each side and multiplying each side by the appropriate reciprocal.*

$$4x+7 = x-5, \; x \in Z \qquad\qquad 2x-3 > 5x-12, \; x \in Z$$
$$\Leftrightarrow 4x-x = -7-5 \qquad\qquad \Leftrightarrow \quad 2x-5x > 3-12$$
$$\Leftrightarrow \quad 3x = -12 \qquad\qquad\qquad \Leftrightarrow \qquad -3x > -9$$
$$x = -4 \qquad\qquad\qquad \Leftrightarrow -\tfrac{1}{3} \times -3x < -\tfrac{1}{3} \times (-9)$$
$$\text{Solution set is } \{-4\} \qquad\qquad \Leftrightarrow \qquad x < 3$$
$$\text{Solution set is } \{x: x < 3, \; x \in Z\}$$

Cumulative Revision Exercises

Exercise A

1. State whether each of the following is true or false.
    - a $3 \in \{\text{prime factors of } 42\}$
    - b $\{0\} = \emptyset$
    - c $\{a, b, a, c\} = \{b, c, d, a\}$
    - d If $A \cap B = A$, then $B \subset A$.

2. List the members of the following sets.
    - a The set of the first six even positive integers.
    - b $\{x: x < 12, \; x \text{ an odd whole number}\}$
    - c $\{\text{common factors of 30 and 42}\}$

3. $S = \{1, 2, 3, 4, 5, 6\}$. Describe S in words. What is $n(S)$? List the set A of even numbers in S, and the set B of odd numbers in S. List $A \cap B$ and $A \cup B$.

4. $E = \{1, 2, 3, \ldots, 10\}$. $A = \{1, 2, 3, 4, 8\}$, $B = \{1, 2, 3, 5, 6, 7\}$; $P = \{4, 5, 6\}$, $Q = \{4, 5, 6, 7, 8, 9\}$; $R = \{1, 2, 3, 4\}$, $S = \{5, 6, 7\}$.
    - a List the elements of $A \cap B$, $P \cap Q$, $R \cap S$, and illustrate each in a Venn diagram.
    - b Which of these is false? $S \subset Q$, $R \subset A$, $P \subset Q$, $B \subset Q$.

5. Make two copies of each of the Venn diagrams in Figure 1, and show by shading $A \cap B$ and $A \cup B$ in each case.

Cumulative Revision Exercises

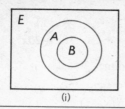
(i)

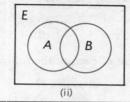
(ii)

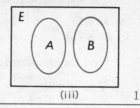
(iii)

6 Given $E = \{1, 2, 3, 4, 5, 6, 7, 8\}$, $A = \{1, 2, 3, 4\}$, $B = \{2, 3, 4, 6, 7\}$, find

    a A' *b* $A \cap B'$ *c* $A' \cap B'$ *d* $(A \cup B)'$

7 Of 114 people interviewed, 62 watched ITV, 70 watched BBC, and 26 watched both ITV and BBC television. Illustrate these results in a Venn diagram. How many of the people did not watch television?

8 P is the solution set of $5x - 2 > 3(x + 4)$, and Q is the solution set of $7x - 9 > 4x + 3$, where $x \in N$. List P, Q and $P \cap Q$.

9 Here the variables are on the set of real numbers.
$A = \{(x, y): y = x + 3\}$, $B = \{(x, y): y \geqslant x + 3\}$,
$C = \{(x, y): 3 \leqslant y \leqslant 6\}$.
On the same diagram, sketch the graphs of A, B, C, and identify the sets $B \cap C$ and $A \cap C$.

10 Find the solution sets of the following open sentences, where the variables are on the set $S = \{-2, -1, 0, 1, 2, 3\}$.

    a $x + 5 = 7$ *b* $x + 2 < 1$ *c* $m + 1 = 0$ *d* x^2 is in S
    e p is negative   *f* p^2 is positive   *g* $0 < z < 3$ *h* $2q \geqslant -2$

11 Given $p = 2$, $q = -1$, $r = 0$, find the values of:

    a $3pqr$ *b* $2p + 2q + 2r$ *c* $pq + qr + rp$
    d $p^2 + q^2 + r^2$ *e* $p^3 + q^3 + r^3$ *f* $2(p + q + r)^2$

12 Simplify the following:

    a $8 + (-7)$ *b* $-5 + 10$ *c* $-13 - 5$
    d $4a - 11a$ *e* $9b + (-11b)$ *f* $8m + (-8m)$
    g $-7 \times (-9)$ *h* $6x \times (-8x)$ *i* $-3 + 7 + 3$
    j $5 - 7 - 8$ *k* $-1 + 3 - 9$ *l* $-2 \times (-2) \times (-2)$
    m $-2x \times (-3x)$ *n* $-7m + 20m - 8m$
    o $-1 \times (-1) \times (-1) \times (-1)$ *p* $20a^2 \div 2a^2$
    q $10b^4 \div (-2b^2)$ *r* $-12c^3 \div -2c^2$

Algebra

13 Use the distributive law to factorize:

a $5p+5q$ b $4n-8$ c $ab+a$ d a^2-4a
e $6p+21q$ f x^2+8x g $10x-20xy$ h $10x-20x^2$

14 Find the simplest form of each of the following:

a $2p+2(p+q)$ b $2p-2(p+q)$ c $2p-2(p-q)$
d $5(x-2)+2(x+5)$ e $x(x+1)-x(x-1)$
f $2a(a+1)-a(a+2)$ g $3(5-2x)+2(7+x)-4(1-x)$

15a Find the sum of $2a-3b$, $3b-4c$ and $-2a+4c$.
    b Subtract $5x-3y+7z$ from $7x+3y-5z$.

16 Simplify:

a (1) $6+(-4)$ (2) $-6+(-4)$ (3) $6\times(-4)$ (4) $-6\div(-4)$
b $(x-2)-(x-3)$ c $(13\times 49)+(13\times 51)$ d $0-x$ e $0\times x$

17 Form an equation for each of the following, then solve it.
   a The sides of a triangle have lengths $8p$ cm, $7p$ cm and $3p$ cm. If the perimeter is 108 cm, find p.
   b The smallest of three consecutive numbers is n. If the sum of the numbers is 225, find the numbers.

18 Find the value of the following when $x=4$, $y=-3$ and $z=2$.
   a $(x+y+z)^2$ b $x^2-y^2-z^2$ c $\frac{1}{2}(3x+4y-5z)$

19 Using the formula $f=\dfrac{v-u}{t}$, calculate f when $v=54$, $u=-18$ and $t=5$.

20 $x\in\{1,2\}$ and $y\in\{4,5\}$. Write down:

a the set of all possible values of $x+y$
b the greatest and least values of x^2+y^2

21 State whether each of the following is true or false.

a $3a^2-1=0$, when $a=-1$ b $x^2+3x+4=0$, when $x=-2$
c $\dfrac{a+b}{a-b}=0$, when $a=-2$ and $b=2$
d $x^2-xy+y^2=37$, when $x=4$ and $y=-3$

22a If $s=ut+\frac{1}{2}ft^2$, find s when $u=48$, $t=\frac{3}{4}$ and $f=-8$.
    b If $s=\frac{1}{2}(u+v)t$, find v when $s=180$, $u=21$, $t=10$.

Cumulative Revision Exercises

23 If $x \in R$, find the solution set of each of the following:
 a $3x - 7 = 26$ b $7x + 23 = 3x - 5$ c $\tfrac{5}{4}x = 20$
 d $3(2x + 7) = 2(5x - 3)$ e $\tfrac{1}{5}(x - 4) = \tfrac{1}{3}(2x - 1)$

24 If $x \in W$, find the solution sets of:
 a $x + 6 > 7$ b $2x - 3 > 4$ c $3x + 10 < 1$
 d $5x + 3 > 2x - 3$ e $8(x - 2) \leqslant 3(x + 3)$

25 A rectangle is $(2x - 5)$ cm long and 3 cm broad. Find x if:
 a its perimeter is 22 cm b its area is 60 cm²

26 The lengths of the sides of a triangle are $9x$ mm, $12x$ mm and $15x$ mm. Show that the triangle is right-angled, and find the length of the hypotenuse when: a $x = 12$ b the perimeter is 144 mm.

27a The sum of two consecutive numbers is 113. Find them.
 b The sum of two consecutive even numbers is 190. Find them.
 c The sum of two consecutive odd numbers is 1728. Find them.

28 $F = \{0{\cdot}25, 0{\cdot}5, 0{\cdot}75\}$. Illustrate the relation *is less than or equal to* on the set F by means of an arrow diagram.

29 $A = \{2, 3, 6, 8, 9\}$. Show by means of an arrow diagram the relation *x is a factor of y* on the set A.
List this relation as a set of ordered pairs.

30 $A = \{1, 5, 9\}$, $B = \{a, b, c, d\}$. Is each of the following true or false?
 a If $R = \{(1, a), (5, c), (9, a)\}$, R is a mapping from A to B.
 b $\{(1, a), (5, b), (9, c), (1, d)\}$ is a mapping from A to B.
 c A and B can be put in one-to-one correspondence.

31a Show in an arrow diagram the relation *is less than or equal to* from the set $A = \{1, 2, 3\}$ to the set $B = \{1, 3, 5\}$. Why is this relation not a mapping?
 b List the set of ordered pairs in the relation *is greater than or equal to* from set A to set B.

32 Make a table for the mapping $x \rightarrow 2x - 1$ from the set $\{0, 1, 2, 3, 4\}$ to the set of whole numbers. Draw the graph of the set of ordered pairs. Hence draw the graph of $x \rightarrow 2x - 1$ on the set of real numbers.

33 Make a table for the mapping $x \rightarrow x^2$ from the set $\{-4, -3, -2, -1, 0, 1, 2, 3, 4\}$ to the set of whole numbers. Draw the graph of the set of ordered pairs. Hence draw the graph of $x \rightarrow x^2$ on the set of real numbers.

Algebra

34 x and y are variables on the set of real numbers. Show by shading the solution sets of the inequations:

   a $y \leqslant 5$ *b* $y > -3$ *c* $y \geqslant x$ *d* $y < x+4$

35 Show by shading the intersection of the solution sets of $x \leqslant 4$ and $y \geqslant -x$.

36 Solve the following systems of equations *graphically* $(x, y \in R)$.

   a $y = x+1$ and $y = -x+3$ *b* $2x+3y = 6$ and $3x-4y = 12$

37 Solve the following, where $x, y \in R$.

   a $\left.\begin{array}{l}x+y = 11\\x-y = 3\end{array}\right\}$ *b* $\left.\begin{array}{l}2x-y = -1\\x+4y = -5\end{array}\right\}$ *c* $\left.\begin{array}{l}3x+4y = 6\\2x-3y = 4\end{array}\right\}$

38 The line $y = 3-x$ and the coordinate axes divide the plane into the seven regions shown in Figure 2.

   a Give the coordinates of the points where the line cuts the axes.
   b State the three regions through which the line $y = x$ passes.
   c In which region is the point $(-3, 4)$?
   d Which regions are defined by $\{(x, y): y \geqslant 0, y \geqslant 3-x\}$?

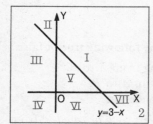
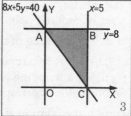
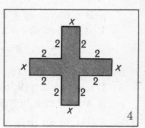

39 Figure 3 shows the lines with equations $x = 5$, $y = 8$, $8x+5y = 40$.

   a Give the coordinates of the points A, B and C.
   b What region is defined by $\{(x, y): x > 0, y > 0, y < \frac{1}{5}(40-8x)\}$?
   c Give a set-builder definition of the shaded region.

40 Make x the subject of the following formulae:

   a $x+m = n$ *b* $px = q$ *c* $ax+b = c$
   d $2(x+c) = d$ *e* $a(x+1) = b$ *f* $(p+q)x = r$
   g $\frac{1}{2}x = k$ *h* $\frac{x}{a} = b$ *i* $\frac{x}{m} = \frac{n}{p}$

Cumulative Revision Exercises

41 Find formulae for the perimeter P m and the area A m² of the cross shown in Figure 4. If $\frac{1}{2} < x < 1$, find corresponding inequations for P and A.

42 $y = ax + b$. If $y = 11$ when $x = 5$ and $y = -10$ when $x = -2$, find a and b.

Exercise B

1   $A = \{1, 2, 3\}$, $B = \{p, q, r\}$. Is each of the following true or false?
   a $3 \in A$ b $p \subset B$ c $A = B$ d $n(A) = n(B)$
   e $A \cap B = \emptyset$ f $A \cup B = \emptyset$ g $A \subset A$ h $\emptyset \subset B$

2   Here the universal set is $W = \{\text{whole numbers}\}$. $S = \{\text{even numbers}\}$ and $T = \{\text{odd numbers}\}$. What is:
   a $S \cap T$ b $S \cup T$ c S' d T'?

3   $E = \{1, 2, 3, \ldots, 12\}$.
   a   List the following subsets of E in full: $A = \{\text{prime factors of 210}\}$, $B = \{\text{odd numbers less than 11}\}$, $C = \{\text{common factors of 24 and 60}\}$.
   b   List the sets $B \cap C$, $A \cap C$ and $A \cap B \cap C$.
   c   If X is a non-empty set such that $X \cap B = \emptyset$ and $X \subset A$, find X.

4   $F = \{0, 1, 2, 5, 7, 9\}$, $G = \{2, 3, 4, 5\}$, $H = \{3, 5, 8, 9\}$. List the sets:
   a $G \cap H$ b $F \cap G$ c $F \cap H$
   d $F \cap (G \cup H)$ e $(F \cap G) \cup (F \cap H)$.

What law is illustrated by the answers to *d* and *e*?

5   There are 120 pupils in a Form III who take one or two or all of Mathematics, French, and Science. All the Science pupils take Mathematics. 108 take Mathematics, 106 take French, and 92 take Science. 88 take French and Science. Draw a Venn diagram to illustrate these statements, and find:
   a   how many take Mathematics but not French or Science;
   b   how many take Mathematics and French but not Science.

6   $E = \{1, 2, 3, 4, 5, 6, 7, 8\}$, $A = \{1, 3, 5, 7\}$, $B = \{2, 4, 6, 8\}$, $C = \{3, 6\}$, and $D = \{3\}$.
Say whether each of the following statements is true or false:

Algebra

    a $A' = B$ b $A \cup C = (B \cap D)'$ c $(A \cap B) \cap C = C$
    d As replacements for X, four different sets satisfy
$$X \cup \{1, 2\} = \{1, 2\}.$$

7   Figure 5 illustrates A = {people who play golf}, B = {people who play cricket}, C = {people who play football} in a certain area. Describe in words the set represented by:

    a the horizontal shading.   b the vertical shading.

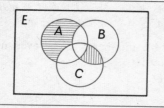
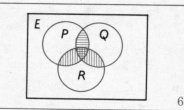

5   6

8   The Venn diagram in Figure 6 shows the relations between sets P, Q, R. State whether each of the following is true or false:

    a The part shaded horizontally represents $P \cup Q$.
    b The part shaded vertically represents $(P \cup Q) \cap R$.
    c $(P \cup Q)' \cap R$ is the empty set.

9   Find the solution sets of the following sentences where the variables are on the set of real numbers:

    a $2x+1 = 4$ b $3x-1 < 13$ c $-3x \geqslant -6$
    d $\tfrac{2}{3}x = 8$ e $2(x-4) = -10$ f $3(5-x) > 8$

10   If $p = 5, q = -2, r = -1, s = 0$, evaluate:

    a $p+q+r+s$ b $p^2+q^2+r^2$ c $3pqrs$
    d $4p-10q$ e $3q+4r$ f $p^3+q^3+r^3$

11   Simplify each of the following:

    a $5x^2+6x^2-x^2$ b $2a-3b+4a$ c $p^2-2q^2-p^2-2q^2$
    d $\tfrac{1}{2}(4x-6y)$ e $4(\tfrac{1}{2}a+\tfrac{3}{4}b)$ f $2(3a-1)-3(2a-1)$
    g $2x(x-5)+x(2x+10)$ h $x^2(x-3)-x(x^2-2x)$

12a Find the sum of $3x+2y, -4x-2y, x+5y$
   b Subtract $2a+3b$ from $3a+2b$, and $-x^2-x$ from $2x^2-3x$

Cumulative Revision Exercises

13 In each of the following state whether answer I, II, or III is correct:

		I	II	III
a	If $xy = 0$, then	x must be zero	y must be zero	Either x or y must be zero
b	If a, b, c are non-zero integers, then	$(ab)c = a(bc)$	$a \div b = b \div a$	$a - b = b - a$
c	$12 - (x - 5)$ is equal to	$7 - x$	$17 + x$	$17 - x$
d	If $a = b+5$ and $b = c-4$ then	$a < c$	$a > c$	$a = c$

14 If $A = 3(x+2y)$ and $B = 2(2x-3y)$ which of the following are correct?

   a $A+B = 7x$ b $A-B = 12y-x$
   c If $B > A$, then $x > 12y$ d If $2A-B = 0$, then $x:y = -1:3$

15a Simplify $3(x+2)-(x-3)$
   b Is it true that $-(a-b) = (-a)+(-b)$?
   c Simplify the following:
      (1) $6+(-3)$ (2) $-6+(-3)$ (3) $6\times(-3)$ (4) $-6\div(-3)$

16 An operation $*$ on the set of integers is defined by
$a * b = ab+2(a+b)$, e.g. $2 * 3 = (2\times 3)+2(2+3) = 6+10 = 16$.
   a Find the values of:
      (1) $1*2$ (2) $2*3$ (3) $(1*2)*3$ (4) $1*(2*3)$
      (5) $(-3*-4)*(-5)$ (6) $-3*(-4*-5)$
   b Is $a*(b*c) = (a*b)*c$ for *all* numbers?

17 Copy and complete the following by placing one of the symbols $<$ or $>$ between the members of each pair:
   a $-2 \ldots 5$ b $-7 \ldots -4$ c $0 \ldots -1$ d $-2 \ldots 2$

18 A square lawn of side 10 m has two rectangular plots x m by 2 m cut from it. If the remaining area is A m², and $A < 80$, show that $x > 5$.
By considering the original length of the lawn, complete $5 < x < \ldots$

19 Use the distributive law to factorize:
   a $4n+6$ b $2ah+2bh+2ch$ c $8x^2-4x$

20 All the pupils in a class of 30 take French or German or both. 22 take French, and 18 take German.

   a If x pupils take French *and* German, how many pupils take
      (1) French only   (2) German only?

Algebra

b By first forming an equation in x, find out how many take French and German.
c Illustrate the problem by means of a Venn diagram.

21 Find the solution sets of the following, where $x \in R$.
  a $3x-5 = 7x+15$ **b** $4(2x-1) = 3(x-5)$
  c $\frac{1}{5}(x-4) - \frac{1}{3}(2x-1) = 0$ **d** $\frac{1}{4}(x+3) - \frac{1}{3}(2x-3) = 2\frac{1}{6}$

22 Find the solution sets of the following, where $x \in R$.
  a $2x+3 < 5x-9$ **b** $5(2x-1) > 2(3x-4)$
  c $16-\frac{1}{3}x \leqslant 2x-5$ **d** $\frac{1}{2}x - \frac{1}{3}(x-4) > \frac{1}{6}(5x-7)$

23 A cuboid is $3x$ cm long, $2x$ cm broad and x cm high. Find formulae for the total length of its edges P cm and its surface area A cm². Calculate x when:
  a $P = 112$ **b** $A = 176$

24 Find the solution sets of the following, where $x \in R$.
  a $6x-8 \geqslant 3x+4$ **b** $\frac{1}{2}(4x-2) < 3$ **c** $\{x: -1 < x+1 \leqslant 4\}$

25 A rectangle is $(3a-6)$ mm long and a mm broad. Given that its perimeter is 60 mm, find its area. Show that $a > 2$.

26 A square of side 10 cm has a circle inscribed in it (i.e. inside the square, touching the sides). A rectangle with sides 6 cm and 8 cm long is inscribed in the circle (i.e. with its vertices on the circle).
  a If the area of the circle is x cm², write down two inequations in x.
  b What is the radius of the circle?
  c Deduce that $1 \cdot 92 < \pi < 4$.

27 Draw an arrow diagram to show the relation *has as factor* from set $A = \{10, 15, 20\}$ to set $B = \{2, 3, 4, 5\}$.

28 State whether each of the following is true or false.
  a Every relation is a mapping.
  b Every mapping is a relation.
  c $(2, 3), (0, -1), (-1, -3)$ are ordered pairs of the mapping $x \to 2x-1$.
  d The graph of the mapping $x \to -10x$, where $x \in Z$, is a straight line through the origin.
  e The set {even numbers between 1 and 11} and the set of working days of the week can be put in one-to-one correspondence.

Cumulative Revision Exercises

29 A mapping from $\{1, 2, 3\}$ to $\{p, q, r\}$ is partly given by $\{(1, p), (2, q), \ldots\}$. Give the possible ordered pairs which would complete the mapping. Which of these do not give a one-to-one correspondence?

30 Show in a table the mapping $x \to 3x - 3$ from $\{-2, -1, 0, 1, 2, 3, 4\}$ to the set of integers. Hence draw a coordinate graph of the mapping from R to R such that $x \to 3x - 3$.

31 Repeat question *30* for the mapping $x \to 3x - x^2$.

32 Sketch the graphs of $A = \{(x, y): y > x\}$ and $B = \{(x, y): y > 2 - x\}$ where $x, y \in R$. Shade the region representing $A \cap B$.

33 Find graphically the solution sets of the systems of equations:
 a $y = 2x - 1$ and $y = -3x + 4$ *b* $x + y = 10$ and $2x - 3y = 10$

34 Solve the following, where $x, y \in R$.
 a $\left.\begin{array}{l} x + y = 0 \\ x - y = -8 \end{array}\right\}$ *b* $\left.\begin{array}{l} 5x - 2y = 17 \\ 2x + 3y = 3 \end{array}\right\}$ *c* $\left.\begin{array}{l} 2x + y = 2 \\ \frac{1}{2}x + \frac{1}{3}y = -1 \end{array}\right\}$
 d $\left.\begin{array}{l} \frac{1}{2}(x-1) + y = 6 \\ \frac{1}{3}(y-1) + x = 6 \end{array}\right\}$ *e* $\frac{1}{4}(x-1) - \frac{1}{3}(y+1) = 0,\ 8x - 7y = 15$

35 Make x the subject of each of the following formulae:
 a $y = 3(2x + 1)$ *b* $y = mx + c$ *c* $a = x^2 - b$
 d $\dfrac{x}{a} = \dfrac{b}{c}$ *e* $\dfrac{x+a}{b} = \dfrac{c}{d}$ *f* $ax + bx = c$

36 Make n the subject of the following formulae:
 a $nE = I(r + Rn)$ *b* $na = \sqrt{(1 - n^2)}$ *c* $d = \dfrac{t(n-1)}{n}$

37 The rectangle and the circle shown in Figure 7 are equal in area and they are between the same parallels. Find a relation between x and r.

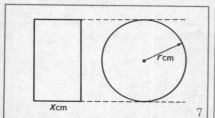

7

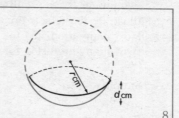

8

Algebra

38 Figure 8 shows a bowl of depth d centimetres, which is part of a hollow sphere of radius r centimetres. The volume, V cubic centimetres, of the bowl is given by the formula $V = \frac{1}{3}\pi d^2(3r-d)$.

    a Make r the subject of the formula.
    b Find r when $V = 17\frac{1}{3}\pi$ and $d = 2$, and hence calculate the radius of the circular top of the bowl.

39 The formula for the total surface area of a cylinder is $A = 2\pi r(r+h)$. Change the subject of the formula to h, then calculate h when $A = 154$ and $r = 3\frac{1}{2}$, taking $\frac{22}{7}$ as an approximation for π.

40 If an electrical circuit has two resistors r_1 and r_2, then the total resistance R in the circuit is related to the resistances r_1 and r_2 by the formula $\frac{1}{R} = \frac{1}{r_1} + \frac{1}{r_2}$. Solve for R in terms of r_1 and r_2, and hence calculate R when $r_1 = 30$ and $r_2 = 45$.

41 *a* $x^2+px+q = 0$ when $x = 5$ and also when $x = -2$. Find p and q.
    b The lines $y = ax+b$ and $y = mx$ intersect at the point P. Find the coordinates of P in terms of the constants a, b and m.

42 Water is flowing through a cylindrical pipe of diameter d cm at the rate of v cm per second. Find a formula for the volume Q cm^3 delivered per minute.

43 Given that the universal set is the set of real numbers, find the complements of the subsets:

$$A = \{x: x \geqslant 1\}, \quad B = \{x: -2 < x < 2\}.$$

44 A club has 50 members and each member is selected for at least one team in football, hockey or swimming. 21 are selected for football, 19 for hockey and 22 for swimming. 7 are listed for both football and swimming. None are listed for both football and hockey.

    a Illustrate these facts on a Venn diagram.
    b How many are listed for both swimming and hockey?
    c How many are listed for swimming only?

45 *a* Simplify: (*1*) $\dfrac{15x-20}{5}$ (*2*) $\dfrac{8a+12}{-4}$ (*3*) $\dfrac{-9x^2-12xy}{3x}$

    b If $A = \frac{1}{2}n-1$ and $B = \frac{1}{3}n+1$, find n such that $4A = B$.

Geometry

Geometry

Note to the Teacher on Chapter 1

Translation is one of the transformations treated explicitly in this geometry course. As in the case of the other transformations—reflection (Book 3), dilatation (Book 6) and rotation (Book 6)—the sophisticated view of translation is that it is a 'point transformation of the plane', i.e. a one-to-one mapping of the set of points of the plane on to itself; under the transformation, to each point of the plane corresponds just one 'image point', and each point is itself the image of just one point. Each transformation may be regarded as a refinement of, and based on abstraction from, real events in the physical world.

Thus we begin in *Sections* 1-2 with the displacements of ships, aircraft, etc. In *Section* 3 we attend to the way such a displacement affects individual features or elements of the object moved. The object moved is sometimes (a model of) the plane itself, e.g. a tracing of a pattern or a tiling. From this experience the idea of translation is extracted.

This development of translation begins with a reminder of geographical bearings already dealt with in Book 1, and leads on to the notion that a displacement is not really defined by bearing alone but requires a magnitude or length. This is developed in *Section* 2 where displacements are represented by directed line segments, which are completely defined by their magnitudes, directions and initial points.

In this chapter we have avoided the difficulty of a symbol for a zero line segment. Later on, we shall have to introduce a symbol for a zero vector. Just as a finite vector, or translation, can be represented by an infinite set of directed line segments, so the zero vector might be represented by $\overrightarrow{AA}$, $\overrightarrow{BB}$, $\overrightarrow{PP}$, etc. This is a complication that does not seem to be necessary at the present stage and the text simply gives $\overrightarrow{AB} \oplus \overrightarrow{BA} = 0$.

The translation of the plane as the translation of all points in the plane needs to be emphasized strongly. The language of mapping and images is used where it appears to be appropriate in the examples.

To pupils familiar with coordinates at an early stage the use of a number-pair notation for expressing translation of the coordinate plane and its application to points with given coordinates should prove a natural development. This notation is also a useful piece of groundwork for the more systematic development of vectors later, and is developed in *Section* 4.

A short *Section* 5 on patterns is introduced to help to emphasize further the translation of *all* points in the plane.

In *Section* 6, the combination of two translations is dealt with in line segment form and in component form. The line segment form has its source in the discussion of displacements in *Section* 2. It should not be forgotten that a directed line segment *represents* a translation; it is not in itself a translation. A statement like $\overrightarrow{AB} \oplus \overrightarrow{BC} = \overrightarrow{AC}$ might be thought of in words as 'The translation represented by $\overrightarrow{AB}$ followed by the translation represented by $\overrightarrow{BC}$ results in the translation represented by $\overrightarrow{AC}$.' To be able to write $\overrightarrow{AB} + \overrightarrow{BC} = \overrightarrow{AC}$ we would need to show that the operation represented by $\oplus$ follows all the same rules as that represented by $+$.

When we come to the component form, the adding of components itself implies that all the rules of addition (commutativity, associativity, etc.) will be followed, and the text drops the symbol $\oplus$ in favour of $+$. The zero symbol here is, of course, $\binom{0}{0}$.

In the more difficult examples some hints are given of a possible algebra of translations, or vectors, but the systematic study is left over to a later chapter.

Translation

1 Position and direction

It is often necessary to describe the position of a place or an object. For example, we may want to
a tell someone where we live
b describe the position of a book in a bookcase or library
c find the position of our town on the surface of the earth
d give the position of a ship at sea or an aircraft in space.

Some methods are based on coordinates of the kind we have already used. Others depend on geographical bearings using the mariner's compass or the three-figure bearings shown in Figures 1 and 2.

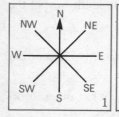
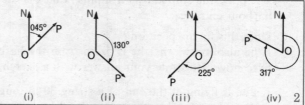

Exercise 1

1. In Figure 2(i), the three-figure bearing of P from O is 045°. Write down the bearing of P from O in (ii), (iii) and (iv).

2. What would be the three-figure bearing of an aircraft leaving an airport and flying:
   a east   b south   c west   d north-east?

3. Show in sketches as in Figure 2 the direction of an object from O if its bearing is:
   a 060°   b 100°   c 280°   d 315°.

Geometry

4. A point B is south-west of A. Make a sketch and give the three-figure bearing of B from A.
   Give the bearing of A from B by:

   a mariner's compass   b three-figure bearing.

5. Choose an origin O near the centre of a page of squared paper. Taking a suitable scale, plot points to mark the positions of the following objects:

   P: 10 km west of O
   Q: 8 km east and 6 km north of O
   R: north-west of O and north of P.

6. Write down information that describes where you live.

7. Give a description of the position of a book in a bookcase.

8. What information can be used to describe the position of your town on a map?

9. a If you wish to describe the position of Edinburgh in relation to London, why is it not sufficient to say how far away Edinburgh is?
   b Describe completely the position of Edinburgh in relation to London.

10. The pilot of a ship knows from his radar that he is south-west of the harbour entrance.

    a Does this fix his position?
    b If he also knows that he is 5 km from the harbour, is his position fixed now? Illustrate your answer with a sketch.

11. A kite is flying at the end of a string 30 m long. What other information is required to fix its position? Make a sketch.

2 Displacements

Introductory Exercise

A pilot flies 200 km on a bearing of 043° from Ayton to Beeton. He then alters course to 087° and flies 350 km to Ceeton.

With the aid of Figure 3 make a scale drawing of his flight, taking a scale of 1 cm to represent 50 km.

Displacements

From your drawing find the distance and bearing of Ceeton from Ayton.

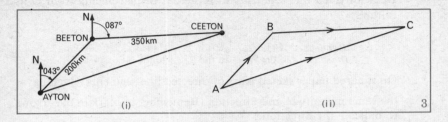

The plane's actual journey or *displacement* from Ayton to Beeton has been *represented* in your drawing by the line AB.

We can draw the line from A to B when we know
(i) the length of AB (ii) the direction from A to B (iii) the position of A.

That is, we have to take account of the *magnitude* and *direction* of AB. We obtain a *directed line segment*, which we write in the form $\vec{AB}$.

Figure 4 shows how the magnitude (or length) and direction of a displacement can be represented by a directed line segment.

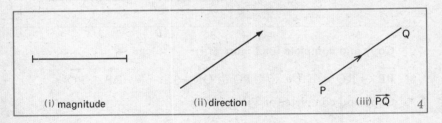

Notice that, in Figure 3, the journey from Ayton to Beeton followed by the journey from Beeton to Ceeton, and the direct journey from Ayton to Ceeton, both take the aircraft from the same starting point to the same destination. In this sense the two journeys are said to be equal, so that using Figure 3(ii), we can write $\vec{AB} \oplus \vec{BC} = \vec{AC}$ where $\oplus$ means 'followed by'.

It would be quite wrong to write AB+BC = AC, for here we would be thinking only of the magnitudes or lengths of AB, BC and AC.

Geometry

Exercise 2

1. Draw directed line segments to represent displacements from O as follows:

	a	b	c	d
Magnitude	50 km	60 km	40 km	70 km
Direction	to the W	to the S	to the NE	085°

2. On squared paper sketch pairs of directed line segments which have:
   a. the same magnitude and direction (Remember to mark in the arrows to indicate the direction.)
   b. the same magnitude, but opposite directions
   c. different magnitudes, but the same direction.

3. a. Why would it be wrong to write AB + BC = AC for Figure 5(i)?
   b. Use the arrow notation for directed line segments, and the symbol ⊕ for 'followed by' to write a correct statement.

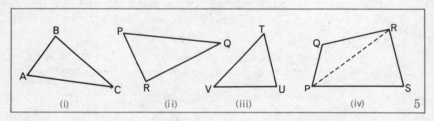

(i) (ii) (iii) (iv) 5

4. Copy and complete for Figure 5(ii):
   a. $\overrightarrow{PR} \oplus \overrightarrow{RQ} = \ldots$ b. $\overrightarrow{PQ} \oplus \overrightarrow{QR} = \ldots$ c. $\overrightarrow{RP} \oplus \overrightarrow{PQ} = \ldots$

5. Copy and complete for Figure 5(iii):
   a. $\overrightarrow{VU} \oplus \overrightarrow{UT} = \ldots$ b. $\overrightarrow{UV} \oplus \ldots = \overrightarrow{UT}$ c. $\ldots \oplus \overrightarrow{TV} = \overrightarrow{UV}$

6. Give two correct statements for journeys from P to R using Figure 5(iv).

7. A car travels 30 km along a road and 30 km back again.
   a. If we describe its first journey by $\overrightarrow{AB}$, how could we describe its return journey?
   b. Copy and complete $\overrightarrow{AB} \oplus \overrightarrow{BA} = \ldots$
   (Note that we should really have a new symbol for a zero displace-

Translation

ment. Since, however, a zero displacement has no direction it is convenient to use the zero number, 0.)
c Would it be true to say AB + BA = 0?

8 A car travels along a triangular route with displacements denoted by $\vec{AB}$, then $\vec{BC}$, then $\vec{CA}$.

a Why is $\vec{AB} \oplus \vec{BC} \oplus \vec{CA} = 0$ true?
b Why is AB + BC + CA = 0 false?

9 a Sketch a route which has four parts and which starts and finishes at the same point.
b Letter it and make some true statements using directed line segments.

3 Translation

Exercise 3

1 A packing case is dragged, without turning, a distance of 8 metres in a straight line. Corner A finishes at A', as shown in Figure 6.

a How far has B moved? Where does it finish?
b How far has C moved? Where does it finish?
c How far has the Trade Mark moved? In what direction?

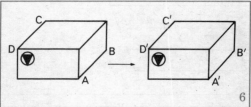
6

7

2 Figure 7 shows the position of a squad of soldiers all facing in the direction of the arrow. The men are ordered to take 8 paces forward.

a Describe the displacement of the man in:
(1) the centre of the front row   (2) the front row and third column from the left.

Geometry

    b On squared paper mark the original positions of all the men, and their positions after obeying the instruction to take 8 paces forward.

    c Comment on the shape and size of the two formations of dots which you have drawn.

3 The distance between two bus stops on a straight road is 300 metres. How far will *a* the driver *b* a back-seat passenger *c* a dent in the side of the bus have travelled between the two stops?

4 In questions *1-3* the magnitude, or size, of the movement was stated. What other information about the displacement was given?

5 Figure 8 shows a tiling of congruent parallelograms with sides 2 cm and 1 cm long. The lines in the diagram are used as guide lines along which tiles may be slid.

    a If tile 6 moves to position 7, how far has each vertex of tile 6 moved? How far has every point on tile 6 moved?

    b Tile 6 moves along the guide lines parallel to BC. Describe the direction of motion of every point on tile 6.

6 *a* If tile 6 moves to position 14, how far has every point on tile 6 moved?

    b If tile 6 moves to position 18, describe the direction of motion of every point on tile 6.

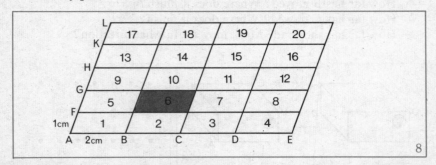

Figure 8

7 Imagine that a tracing has been made of Figure 8 and lettered A′, B′, C′, ..., L′ to correspond to A, B, C, ..., L on the tiling. Suppose the tracing slides over the tiling so that the tracing of tile 6 covers tile 7.

    a Which tiles will be covered by the tracing of tiles 1, 11, 15 and 18?

    b Which letters will be found under the letters A′, B′, C′ and D′ on the tracing?

Translation

8 Suppose the tracing slides over the tiling so that the tracing of tile 6 covers tile 14.

a Which tiles will be covered by the tracings of tiles 2, 8, 11 and 5?
b Which letters will be found under the letters A', F', G' and H' on the tracing?

In Section 2 we used the word *displacement* to describe the movement of an object through a certain distance in a certain direction. We represented the displacement in magnitude and direction by a directed line segment.

We call a displacement of all points in the plane a *translation*.

So we can represent the translation by a directed line segment. For example, the translation of the plane, i.e. of all points in the plane, which takes tile 6 to position 7 in Figure 8, can be represented by $\vec{AB}$, or $\vec{BC}$, or $\vec{CD}$, or $\vec{DE}$. The translation which takes tile 6 to position 14 can be represented by $\vec{AG}$, or $\vec{FH}$, or $\vec{GK}$, or $\vec{HL}$.

Under a certain translation A → A' and B → B' as shown in Figure 9. It follows that $\vec{AA'} = \vec{BB'}$ and hence AA' = BB' and AA' ∥ BB'.

Also $\vec{AB} = \vec{A'B'}$ and hence AB = A'B' and AB ∥ A'B'.
This makes ABB'A' a parallelogram.

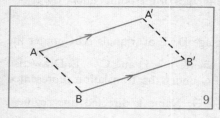

9

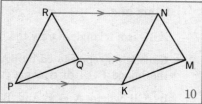

10

Exercise 4

1 AB = 4 cm. Under a translation of length 6 cm, A → D and B → C.
a What are the lengths of AD, BC and DC?
b Name two pairs of parallel lines in your figure.

2 PQR is a triangle with PQ = 5 cm, QR = 6 cm and RP = 7 cm. Under a translation of length 8 cm, P → P', Q → Q' and R → R'.

Geometry

What are the lengths of:

a PP' *b* RR' *c* Q'R' *d* R'P'?

3 In Figure 10 a translation maps P to K, Q to M and R to N, so that K, M and N are the *images* of P, Q and R respectively.

a Name a set of three parallel straight lines in the diagram.
b Name three equal lines in the diagram.
c Name the image of triangle PQR. These two triangles are .... (What is the missing word?)
d Name three parallelograms in the diagram.

4 Sketch a square ABCD. A translation maps A to C.

a Draw the image square CXYZ under this translation.
b Name three directed line segments equal to $\overrightarrow{AC}$.

5 Draw a rectangle ABCD and its image AB'C'D' under a half turn about A. What translation would map ABCD to C'D'AB'?

6 A circle has centre O and A is a point on the circumference.

a Draw the image of the circle under the translation represented by $\overrightarrow{OA}$.
b Take any points P, Q and R on the given circle and, by drawing directed line segments, find the images of the points under the translation in *a*.

7 ABC is a triangle. Draw the image DAE of triangle ABC under the translation represented by $\overrightarrow{BA}$ where A → D and C → E. Draw also the image FGB of triangle ABC under the translation represented by $\overrightarrow{CB}$ where A → F and B → G. Name a directed line segment which represents the translation of:

a the first image to the second *b* the second image to the first.

8 State whether each of the following is true or false. When a figure has a translation applied to it:

a all points move the same distance
b not all points move in the same direction
c all lengths in the figure remain unchanged
d at least one point remains fixed (invariant)

Translation

e the translation can be represented by a directed line segment
f the figure and its image are congruent.

Exercise 4B

1 Figure 11 shows part of a parallelogram tiling.
  a Using the given letters, name four congruent triangles in the figure.
  b Name which of these triangles can be mapped to triangle AFE by a translation, and state a representative of the translation in each case.
  c Which of the four triangles cannot be mapped to triangle AFE by a translation? State another transformation that could be used to effect the mapping.

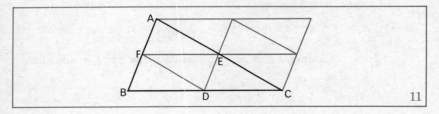

2 Copy Figure 11. Sketch the image of triangle ABC under a translation represented by $\overrightarrow{BD}$. Name two other directed line segments in the figure that would represent the same translation.
   If A′ is the image of A under this translation, define the translation that maps triangle FED to triangle AA′E.

3 PQRS is a parallelogram and P′ is a point on the diagonal PR. Sketch the image P′Q′R′S′ of PQRS under a translation represented by $\overrightarrow{PP'}$. Join QQ′, RR′ and SS′.
   Name as many lines as you can which are equal and parallel to PP′. How many parallelograms can you see in the whole figure?

4 Draw a triangle ABC (near the centre of a page, and not too large). Sketch the images of triangle ABC under translations represented by $\overrightarrow{BC}$, $2\overrightarrow{BC}$, $\overrightarrow{CB}$ and $2\overrightarrow{CB}$.
   Sketch now the images of the whole diagram under the translations represented by $\overrightarrow{BA}$ and $\overrightarrow{AB}$.

If you were to continue these processes indefinitely, what statement could you make about triangle ABC and a tiling of the whole plane?

4 A number pair notation

A very useful way to describe a translation is by means of a *number pair*. Suppose that the translation is represented by any one of the *equivalent* directed line segments in Figure 12.

The effect of the translation is to map each point of the plane to another point by going '1 to the right and 3 up'. Instead of writing '1 to the right and 3 up' it is usual to write the translation as $\binom{1}{3}$.

This is a number pair in which 1 and 3 are called the *components* of the translation.

Using components it is easy to translate any point in coordinate terms.

Example 1. Under the translation $\binom{1}{3}$ the point (2, 4) becomes (2+1, 4+3), i.e. (3, 7).

Example 2. Under the translation $\binom{-2}{2}$ the point (3, −3) becomes (3+(−2), −3+2), i.e. (1, −1).

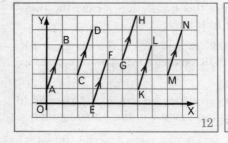
12

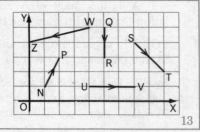
13

Exercise 5

1 Why could each of the directed line segments shown in Figure 12 be used to describe the same translation?

A number pair notation

2   Name the images of C, G and K under the translation represented in Figure 12.

3   Give the coordinates of the images of the following points under the translation represented in Figure 12: (2, 0), (3, 3), (7, 1), (−4, 0), (−1, −1).

4   Express in number pair form the translations represented by the directed line segments $\vec{NP}$, $\vec{QR}$, $\vec{ST}$, $\vec{UV}$ and $\vec{WZ}$ in Figure 13.

5   A translation is described by the number pair $\begin{pmatrix} 2 \\ -1 \end{pmatrix}$. Write down the coordinates of the images of the points (0, 0), (5, 7), (−2, −2), (1, 1) and (10, 10) under this translation.

6   Under a certain translation, A (3, 4) → A′ (2, 8).
  a Represent this translation by a number pair.
  b What are the images of (1, 2), (6, 6) and (1, −4) under this translation?

7   Copy Figure 14. Draw the image of each shape under the translation $\begin{pmatrix} -2 \\ 3 \end{pmatrix}$.

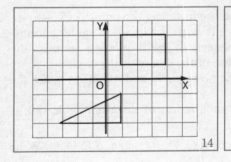
14

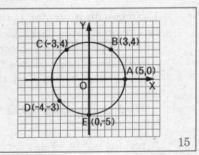
15

8   On squared paper copy the circle shown in Figure 15. Draw the image of this circle under a translation $\begin{pmatrix} 6 \\ 5 \end{pmatrix}$. Mark the images A′, B′, C′, D′ and E′ of the points A, B, C, D and E on the circumference of the original circle. Write down the coordinates of A′, B′, C′, D′ and E′. Write down also the coordinates of the centre of the image circle.

Geometry

9 Under the translation $\begin{pmatrix} 5 \\ -2 \end{pmatrix}$, the point A (4, 4) maps to A′, and B (0, 2) maps to B′.

   a Find the coordinates of A′ and B′. Show in a diagram the points A, B, A′ and B′.
   b Name two equal directed line segments and a parallelogram in your diagram.

10 Under a translation $\begin{pmatrix} -3 \\ 1 \end{pmatrix}$, A′ (6, 2) is the image of A and B (−2, −3) has an image B′.

   a Find the coordinates of A and B′ and illustrate on squared paper.
   b Name two pairs of equal directed line segments in your diagram.

Exercise 5B

1 The coordinates of the vertices of an isosceles triangle are P (0, 6), Q (−1, 0) and R (1, 0). Draw the triangle on squared paper and show its images under the translations represented by

   a $\overrightarrow{QR}$ *b* $\overrightarrow{RQ}$ *c* $\overrightarrow{OP}$, where O is the origin.

In each case, state the components of the translation and the co-ordinates of the new vertices.

2 A circle with its centre at the origin O has the point A (5, −6) on its circumference. If A′ (12, −7) is the image of A under a translation, write down the components of the translation and also the coordinates of the new position of the centre of the circle.

On squared paper draw the given circle and its image under the above translation.

What can you state about each radius of the given circle and its image radius?

3 A rectangle ABCD has vertices A (3, 2), B (9, 2) and D (3, 6). Write down the coordinates of C.

State the coordinates of the images of all four vertices under a translation represented by $\tfrac{1}{2}\overrightarrow{AC}$.

4 On squared paper make a drawing of a chess-board (64 squares). Number the bottom row 1 to 8 from left to right, the next 9 to 16 and so on up to 64. Use the bottom edge as the *x*-axis and the left-hand edge as the *y*-axis.

Patterns

101

State the components of the translations that will map square:

- *a* 1 to 32
- *b* 3 to 57
- *c* 33 to 7
- *d* 64 to 18
- *e* 10 to 42
- *f* 24 to 18

5 $\binom{3}{2}$ is a translation in component form.

- *a* Write down the components of a translation that will displace all points in the plane.
  - (*1*) twice as far in the same direction
  - (*2*) three times as far in the same direction
  - (*3*) ten times as far in the same direction
  - (*4*) five times as far in the opposite direction.
- *b* Repeat the question starting with the components $\binom{a}{b}$.
- *c* Suggest a rule for multiplying a translation in component form by a number.

---

5 Patterns

Exercise 6

1 a Mark the origin O near the centre of a page of squared paper. Plot the points (1, 1), (4, 0), (1, −1), (2, 0) and join them up to make a kite shape.
- *b* Draw this shape again, moving each point 4 units to the right.
- *c* Repeat, each time moving points 4 units to the right, until you come to the edge of the page.
- *d* Extend the pattern to the left in the same way.
    You now have a regular *strip pattern*.

2 Think of the pattern extended as far to the left and right as anyone could possibly imagine (see Figure 16).

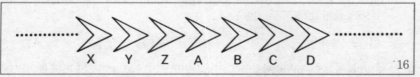

16

Make (or imagine) a tracing of the pattern, and let the shapes ... Z', A', B', C', ... be the tracings of shapes ... Z, A, B, C ....

Geometry

102

    a If A′ is slid on to B, what shape is under (*1*) X′ (*2*) Y′ (*3*) Z′ (*4*) B′?
    b Is there a traced shape which will fit over X?
    c Is there a traced shape which will fit over the tenth shape to the left of A? the millionth shape to the right of A?

3   Make some other strip patterns.

4   Figure 17 shows part of a mathematical pattern which is repeated as far to the left and the right as anyone can imagine.

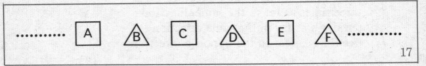

17

    Which of the translations represented by the following would carry the pattern on to itself again?

  a $\overrightarrow{AB}$ *b* $\overrightarrow{AC}$ *c* $\overrightarrow{BD}$ *d* $\overrightarrow{AD}$ *e* $\overrightarrow{BC}$ *f* $\overrightarrow{CD}$ *g* $\overrightarrow{AE}$ *h* $\overrightarrow{FB}$

5   Figure 18 shows part of a regularly repeating pattern. Copy it on squared paper and continue the pattern in all directions as far as your page will allow. Imagine it extended beyond your page in all directions.

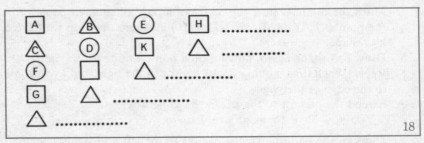

18

    Make, or imagine, a tracing of the pattern. Which of the translations represented by the following would carry the pattern on to itself again (conserve the pattern)?

  a $\overrightarrow{AB}$ *b* $\overrightarrow{EA}$ *c* $\overrightarrow{CB}$ *d* $\overrightarrow{EF}$ *e* $\overrightarrow{FE}$ *f* $\overrightarrow{AG}$ *g* $\overrightarrow{AH}$

6   Think of, or examine, other situations where you might find repeated patterns; for example:

    Wallpaper, curtain material, carpets, tiled walls.

6 Two successive translations

In Figure 19(i), $\vec{AB}$ represents a translation $\begin{pmatrix}3\\1\end{pmatrix}$ and $\vec{BC}$ represents a translation $\begin{pmatrix}2\\4\end{pmatrix}$. It can be seen that the translation $\begin{pmatrix}3\\1\end{pmatrix}$ followed by the translation $\begin{pmatrix}2\\4\end{pmatrix}$ is equivalent to a single translation represented by $\vec{AC}$, i.e. $\begin{pmatrix}3\\1\end{pmatrix}$ followed by $\begin{pmatrix}2\\4\end{pmatrix}$ gives $\begin{pmatrix}5\\5\end{pmatrix}$.

We can think of the components separately:
    3 to the right and 2 to the right gives 5 to the right,
    1 up and 4 up gives 5 up;
i.e. we add the components to get the components of the answer.

Similarly in Figure 19(ii) $\vec{DE}$ represents the translation $\begin{pmatrix}3\\2\end{pmatrix}$ and $\vec{EF}$ represents the translation $\begin{pmatrix}-2\\4\end{pmatrix}$.

$\vec{DE}$ followed by $\vec{EF}$ gives $\vec{DF}$;

$\begin{pmatrix}3\\2\end{pmatrix}$ followed by $\begin{pmatrix}-2\\4\end{pmatrix}$ gives $\begin{pmatrix}3-2\\2+4\end{pmatrix} = \begin{pmatrix}1\\6\end{pmatrix}$.

In Figure 19(iii), $\vec{PQ}$ followed by $\vec{QR}$ gives $\vec{PR}$;

$\begin{pmatrix}-3\\-3\end{pmatrix}$ followed by $\begin{pmatrix}6\\-2\end{pmatrix}$ gives $\begin{pmatrix}-3+6\\-3-2\end{pmatrix} = \begin{pmatrix}3\\-5\end{pmatrix}$.

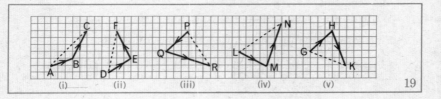

(i)   (ii)   (iii)   (iv)   (v)   19

Geometry

Since we get the answer by adding components it is usual to think of *adding* translations in number pair form, and to use + rather than ⊕, e.g.

$$\binom{3}{1}+\binom{2}{4}=\binom{5}{5}; \quad \binom{3}{2}+\binom{-2}{4}=\binom{1}{6}; \quad \binom{-3}{-3}+\binom{6}{-2}=\binom{3}{-5}.$$

Note: A single translation could be regarded as the sum of two translations parallel to the axes, e.g.

$$\binom{3}{0}+\binom{0}{1}=\binom{3}{1}.$$

Exercise 7

1 For Figure 19, find two translations parallel to the axes which can be added to give the translations represented by

    a $\vec{BC}$ *b* $\vec{DE}$ *c* $\vec{EF}$ *d* $\vec{PR}$ *e* $\vec{LM}$.

2 State your answers to question *1* in the form of equations in number pair form, e.g. $\binom{1}{0}+\binom{0}{4}=\binom{1}{4}$.

3 In Figure 19(i) we have $\vec{AB} \oplus \vec{BC} = \vec{AC}$ and $\binom{3}{1}+\binom{2}{4}=\binom{5}{5}$.

Write down similar equations for   *a* Figure 19(ii)

   b Figure 19(iii)   *c* Figure 19(iv)   *d* Figure 19(v)

4 *a* On squared paper draw directed line segments $\vec{UV}$ and $\vec{VW}$ to represent the translations $\binom{-2}{4}$ and $\binom{3}{-1}$.

   b From your diagram write down a true equation containing directed line segments.

   c Write down the corresponding equation in number pairs.

5 Find a single number pair for each of the following:

    a $\binom{1}{2}+\binom{-2}{3}$ *b* $\binom{1}{2}+\binom{-1}{-2}$ *c* $\binom{6}{-5}+\binom{5}{-6}$ *d* $\binom{p}{q}+\binom{r}{s}$

6 The translation $\binom{3}{2}$ followed by the translation $\binom{a}{b}$ gives the translation $\binom{5}{6}$. What are the values of *a* and *b*?

Two successive translations

7. A cloud of smoke represented by a set of points in space is blown 40 m north and then 30 m east. Show in a sketch the translation of the smoke particles and indicate the shortest path which would have taken them to the same final position.

8. A tidal flow of water representing a translation of points in water moves 25 km south and then 15 km south-west. Show in a sketch the translation of the water and indicate the shortest route which would have taken the water to the same final position.

9. A captain has to sail his ship from a port A to a port B which is 120 km from A on a bearing 058°. Owing to the shape of the coastline he must first sail 70 km to C on a bearing 082° and then make directly for B. From an accurate scale drawing measure the distance and bearing of B from C.
   Write a true statement about directed line segments in triangle ABC.

10. PQR is a triangular route on a map.
    In what way is the journey represented by $\overrightarrow{PR}$ equivalent to the journey represented by $\overrightarrow{PQ} \oplus \overrightarrow{QR}$?
    Discuss possible advantages and disadvantages of the two routes if you travel:
    a by motor-car   *b* by ship   *c* by plane.

We have seen that translations can be 'added' when they are represented by 'nose-to-tail' directed line segments as in Figure 19. Suppose we wish to combine the translations represented by $\overrightarrow{AB}$ and $\overrightarrow{PQ}$ in Figure 20. The translation represented by $\overrightarrow{PQ}$ can be represented by any directed line segment with the same magnitude and direction as $\overrightarrow{PQ}$. Some of these are shown in colour in Figure 20.

If we choose $\overrightarrow{BC}$ we can say that the translation represented by $\overrightarrow{AB}$ followed by the translation represented by $\overrightarrow{PQ}$ gives the same

Geometry

result as the translation represented by $\vec{AB}$ followed by the translation represented by $\vec{BC}$. We *choose* $\vec{BC}$ because $\vec{AB}$ and $\vec{BC}$ fit nose-to-tail.

Hence $\vec{AB} \oplus \vec{PQ} = \vec{AB} \oplus \vec{BC} = \vec{AC}$.

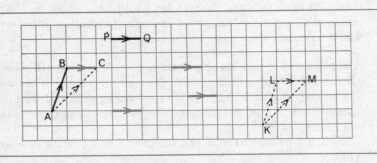

If we care we can choose *any* representatives of the two translations provided they fit nose-to-tail, e.g. $\vec{KL}$ and $\vec{LM}$ in Figure 20.

Under the two successive translations we have $\vec{KL} \oplus \vec{LM} = \vec{KM}$, or in number pair notation $\begin{pmatrix}1\\3\end{pmatrix} + \begin{pmatrix}2\\0\end{pmatrix} = \begin{pmatrix}3\\3\end{pmatrix}$.

If K is the point (10, 2) then M is (10+3, 2+3), i.e. (13, 5).

The *composition* of two translations is achieved by means of a nose-to-tail placing of the directed line segments in a triangle.

Exercise 8

1. On squared paper mark points P (1, 2) and Q (3, 7). Join PQ. On your diagram draw five directed line segments each representing the same translation as $\vec{PQ}$.

2. Repeat question *1* when P is (5, 2) and Q is (1, 6).

3. Repeat question *1* when P is (3, −1) and Q is (−2, 1).

4. *a* Copy Figure 21 and by choosing a suitable equivalent directed line segment for $\vec{FG}$ illustrate the result of $\vec{OE} \oplus \vec{FG}$.
   b Write down an equation in number pairs for the combination carried out in *a*.

Two successive translations

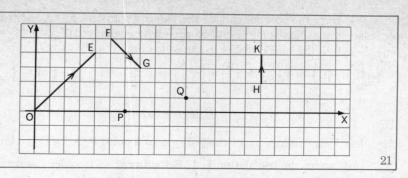

21

5 a On your diagram for question **4** mark the three images of P under the translations $\overrightarrow{OE}$, $\overrightarrow{FG}$ and $\overrightarrow{OE} \oplus \overrightarrow{FG}$. Repeat for Q.
  b Write down the coordinates of these six images.

6 By using equivalent directed line segments, draw a directed line segment which is equal to:

  a $\overrightarrow{OE} \oplus \overrightarrow{HK}$ **b** $\overrightarrow{FG} \oplus \overrightarrow{HK}$ **c** $\overrightarrow{HK} \oplus \overrightarrow{FG}$.

7 a Explain what happens when you try to represent $\overrightarrow{FG} \oplus \overrightarrow{GF}$.
  b Show this result in terms of number pairs.

8 With the help of Figure 22, draw directed line segments on squared paper to represent the following:

  a $\overrightarrow{AB} \oplus \overrightarrow{CD}$ **b** $\overrightarrow{CD} \oplus \overrightarrow{EF}$ **c** $\overrightarrow{EF} \oplus \overrightarrow{GH}$
  d $\overrightarrow{CD} \oplus \overrightarrow{GH}$ **e** $\overrightarrow{KL} \oplus \overrightarrow{AB}$ **f** $\overrightarrow{AB} \oplus \overrightarrow{GH}$

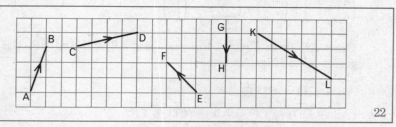

22

9 One translation is given by $\begin{pmatrix} 4 \\ 3 \end{pmatrix}$ and another by $\begin{pmatrix} 3 \\ -3 \end{pmatrix}$.

  a Write down the images of the point A (2, 3) under each translation.
  b What single number pair defines $\begin{pmatrix} 4 \\ 3 \end{pmatrix}$ followed by $\begin{pmatrix} 3 \\ -3 \end{pmatrix}$?

Geometry

10 What are the components of the identity translation (i.e. the translation that leaves all points on the plane unchanged)?
Fill in the blank spaces in the following.

a $\begin{pmatrix}2\\3\end{pmatrix}+\begin{pmatrix}\\\end{pmatrix}=\begin{pmatrix}0\\0\end{pmatrix}$ b $\begin{pmatrix}0\\0\end{pmatrix}+\begin{pmatrix}5\\-1\end{pmatrix}=\begin{pmatrix}\\\end{pmatrix}$ c $\begin{pmatrix}-2\\-4\end{pmatrix}+\begin{pmatrix}\\\end{pmatrix}=\begin{pmatrix}-2\\-4\end{pmatrix}$

Exercise 8B

1 Mark on squared paper the position of a point P (3, 2), Q the image of P under the translation $\begin{pmatrix}4\\1\end{pmatrix}$, R the image of P under the translation $\begin{pmatrix}2\\5\end{pmatrix}$ and S the image of P under the translation $\begin{pmatrix}4\\1\end{pmatrix}+\begin{pmatrix}2\\5\end{pmatrix}$.
What kind of figure is PQSR?

2 A is the point (−1, 3), B is (−3, −3), C is (4, 4) and D is (1, −5).
 a By means of a diagram on squared paper with the origin near the middle of the page, draw a directed line segment equal to $\vec{AB} \oplus \vec{CD}$.
 b Express the translation given by $\vec{AB} \oplus \vec{CD}$ in component form.

3 Repeat question 2a and b in the case of $\vec{AB} \oplus \vec{DC}$.

4 Fill in the missing components in the following:

a $\begin{pmatrix}9\\-4\end{pmatrix}+\begin{pmatrix}8\\2\end{pmatrix}+\begin{pmatrix}-12\\1\end{pmatrix}=\begin{pmatrix}\\\end{pmatrix}$

b $\begin{pmatrix}1\\-1\end{pmatrix}+\begin{pmatrix}-2\\3\end{pmatrix}+\begin{pmatrix}4\\3\end{pmatrix}+\begin{pmatrix}\\\end{pmatrix}=\begin{pmatrix}0\\0\end{pmatrix}$

c $\begin{pmatrix}4\\6\end{pmatrix}+\begin{pmatrix}\\\end{pmatrix}+\begin{pmatrix}-3\\2\end{pmatrix}+\begin{pmatrix}9\\-4\end{pmatrix}=\begin{pmatrix}12\\12\end{pmatrix}$

5 Draw any three directed line segments $\vec{AB}$, $\vec{CD}$ and $\vec{EF}$. Use diagrams to investigate whether:

 a $\vec{AB} \oplus \vec{CD} = \vec{CD} \oplus \vec{AB}$
 b $(\vec{AB} \oplus \vec{CD}) \oplus \vec{EF} = \vec{AB} \oplus (\vec{CD} \oplus \vec{EF})$
 c the operation $\oplus$ has an identity element.
Which laws are you investigating in *a* and *b*?

Summary

1 Displacement

A displacement of an object from a given point A to a point B is defined by its *magnitude* and *direction*.

The displacement can be represented by the directed line segment $\vec{AB}$.

2 Translation

A *translation* is a displacement of all points in the plane through the same distance, in the same direction.

The translation may be represented by any one of a set of directed line segments with the same magnitude and direction.

3 Components

In the coordinate plane, a translation may be defined by its *components*, e.g. a translation $\binom{a}{b}$ moves every point a units parallel to the x-axis and b units parallel to the y-axis.

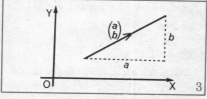

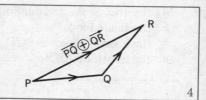

4 Composition of translations

a If $\vec{PQ}$ and $\vec{QR}$ represent two translations, then $\vec{PR}$ represents the combined translation $\vec{PQ} \oplus \vec{QR}$ ($\vec{PQ}$ followed by $\vec{QR}$).

Note. The line segments representing the translations must fit together nose-to-tail.

b In components, $\binom{a}{b} + \binom{c}{d} = \binom{a+c}{b+d}$.

The Calculation of Distance

1 Calculating the length of a line

Exercise 1

1. Figure 1 shows squares with sides of lengths 1, 2, 3, 4 and 5 units. State the area of each square in square units.

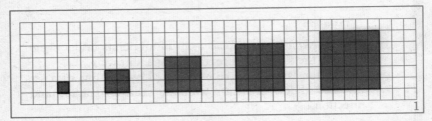

2. What would be the lengths of the sides of squares with the following areas in square units? Use square root tables (pages 266-269) where necessary.
   a 36   *b* 64   *c* 100   *d* 144   *e* 20   *f* 55   *g* 70   *h* 6

3. In Figure 2(i) calculate:
   a the area of the surrounding square, shown by broken lines
   b the shaded area
   c the area of square ABCD
   d the length of AB, to three significant figures.

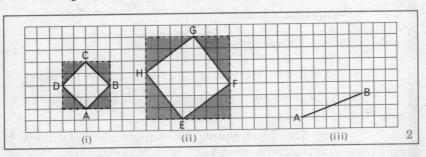

Note to the Teacher on Chapter 2

Pythagoras' theorem is arrived at in this chapter by an argument involving areas, but it is important to emphasize that essentially the theorem is about the calculation of lengths on the Euclidean plane. It seems appropriate as usual to proceed from numerical instances to the general case. Teachers may wish to consider the rotational symmetry of Figure 2(ii) as a class exercise to emphasize that EFGH is a square, even although rotation in general has not been discussed.

In the actual demonstration of the theorem, an attempt has been made to use the simplest possible approach. Probably cut-out shapes would help, as the four right-angled triangles could then be moved into their new positions without rotation or turning over (see Figure 4).

In calculations, statements like $AB^2 = BC^2 + CA^2$ are better avoided. Calculations are about numbers. If the lengths of the sides AB, BC and CA are 5 cm, a cm and 2 cm we are entitled to write $5^2 = a^2 + 2^2$, and to conclude that $a = \sqrt{21}$ and BC $= \sqrt{21}$ cm.

In view of the numerical approach to calculating of length in *Section* 1, it seems to be reasonable to proceed to the Distance Formula in *Section* 3. The fact that $d^2 = \sqrt{[(x_2-x_1)^2 + (y_2-y_1)^2]}$ for *all* positions of (x_1, y_1) and (x_2, y_2) is rather glossed over in the text, but for good pupils the general operation should not be too difficult, if regarded as subtraction on two number lines.

The converse of the theorem has probably a longer history than the theorem itself, at least in particular cases, and it seems to be worth while dealing with this important converse along with the theorem. Attempts to provide integral solutions of the equation $x^2 + y^2 = z^2$ go right back to the time of Pythagoras, and the Topic to Explore glances at the general solution without introducing algebraic difficulties like squaring $m^2 + n^2$ (which will be included in Algebra Chapter 1 of Book 5).

The examples in this chapter have been compiled on the assumption that any pupil who tackles Exercises 1-6 with reasonable success has mastered the essentials. The B examples provide an additional challenge to the better pupils, and as a consequence they contain some quite difficult questions.

(*facing page* 110)

Calculating the length of a line

4 Repeat question *3* for Figure 2(ii) in order to calculate the length of EF, a side of square EFGH.

5 We could describe AB in Figure 2(i) as '2 right, 2 up', BC as '2 left, 2 up' and so on. Describe in the same way EF, FG, GH and HE.

6 a On squared paper draw a line AB which goes '5 right' and '2 up' as in Figure 2(iii). Draw BC and AD going '2 left' and '5 up'.
 b Join DC and describe how you go from D to C.
 c What kind of shape is ABCD?

7 Repeat question *6* where AB goes '3 right' and '5 up', and BC and AD go '5 left' and '3 up'.

8 a On squared paper draw the line IJ '3 right' and '2 up'.
 b Draw the square on IJ, IJKL, as in questions *6* and *7*, and also the surrounding square with its sides on the squared grid.
 c Calculate the area of the square IJKL as in question *3*, and use square root tables to find the length of IJ to 3 significant figures.

9 Repeat question *8* in the case of :
 a square MNOP, where MN is '8 right' and '6 up'
 b square QRST, where QR is '2 right' and '1 up'
 c square UVWX, where UV is '5 right' and '4 up'.

We can use this method to calculate the lengths of lines in the coordinate plane.

Figure 3 shows certain right-angled triangles from the diagrams in questions *3*, *4*, *8* and *9*.

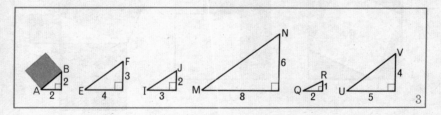

The side opposite the right angle in a right-angled triangle is called the *hypotenuse*, e.g. AB is the hypotenuse of the first triangle.

10a Copy and complete this table, using Figure 3 and your working for questions *3*, *4*, *8* and *9*.

Geometry

Area of square on hypotenuse	$AB^2 = 8$	$EF^2 =$	$IJ^2 =$	$MN^2 =$	$QR^2 =$	$UV^2 =$
Sum of areas of squares on other two sides	$2^2 + 2^2$ $= 4+4$ $= 8$	$4^2 + 3^2$ $=$				

b What do you notice in the table about the two answers in each column?

Note: We have here some instances of a general result about right-angled triangles which has been used by mathematicians and surveyors and others concerned with measurement for 3000 years. It is referred to as 'Pythagoras' theorem' in honour of a Greek mathematician who lived in the sixth century B.C. and gave a proof of its truth.

The theorem states:

In a right-angled triangle the square on the hypotenuse (the side opposite the right angle) is equal to the sum of the squares on the other two sides.

We give a proof of this result in the next section.

2 Pythagoras' theorem

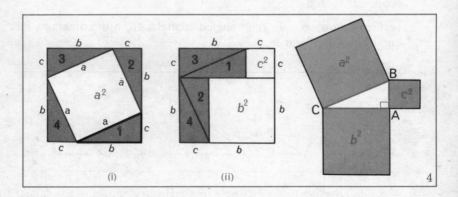

In Figure 4(i), a right-angled triangle has its sides a, b and c units long respectively. The square on $b+c$ is built up as shown. (Compare with the numerical examples in Section 1.)

Pythagoras' theorem

In Figure 4(ii), the square on $b+c$ is built up from b^2, c^2 and the four shaded right-angled triangles from Figure 4(i).

Since the shaded areas in Figures 4(i) and 4(ii) are equal, the unshaded areas are also equal.

Hence $a^2 = b^2 + c^2$.

We have now proved Pythagoras' theorem. This theorem is usually stated as follows:

The square on the hypotenuse of a right-angled triangle is equal to the sum of the squares on the other two sides.

The result is illustrated in the third diagram in Figure 4 in the form:

In triangle ABC, if angle A is a right angle, $a^2 = b^2 + c^2$.

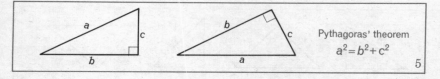

Pythagoras' theorem
$a^2 = b^2 + c^2$

5

Note: We have proved Pythagoras' theorem by means of areas, but we use it mainly to calculate lengths. Tables of squares and square roots are useful in this work.

Example. If in one of the triangles in Figure 5, $b = 9$ and $c = 5$, calculate a.

$$a^2 = b^2 + c^2$$
$$= 9^2 + 5^2$$
$$= 81 + 25$$
$$= 106$$

Then $a = \sqrt{106}$
$= \sqrt{(1 \cdot 06 \times 100)}$
$= 1 \cdot 03 \times 10$
$= 10 \cdot 3$

Exercise 2

1 Use Pythagoras' theorem to write down equations about the lengths of the sides of the right-angled triangles in Figure 6.

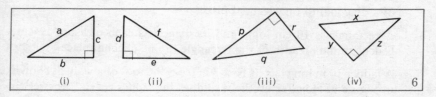

(i)   (ii)   (iii)   (iv)   6

Geometry

2   Use Pythagoras' theorem to calculate x for each triangle in Figure 7.

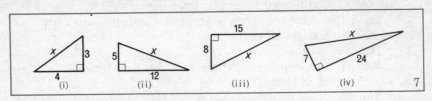

3   *a* The 'addition' statement $a^2 = b^2 + c^2$ can be written in two different ways as a 'subtraction' statement. Copy and complete:
$$b^2 + c^2 = a^2, \text{ so } b^2 = \ldots - \ldots, \text{ and } c^2 = \ldots - \ldots.$$
  b Use these results to calculate x for each triangle in Figure 8.

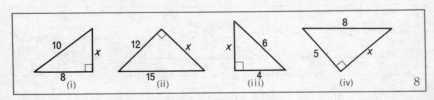

4   The sides of a right-angled triangle measure a, b and c units, a units being the length of the hypotenuse. Copy and complete the following table giving answers to three significant figures where necessary.

a			13	10	17	9		
b	6	2	5	4			40	120
c	8	3			15	6	30	160

Some 'Pythagorean triples' are worth remembering. You may have already noticed the following sets:
    3, 4, 5 and multiples like 6, 8, 10 and 9, 12, 15; 5, 12, 13; 8, 15, 17; 7, 24, 25. The *Topic to Explore* on page 127 investigates this further.

5   The sides of a rectangle are 5 cm and 6 cm long. Calculate the length of a diagonal to one decimal place.

6   A rectangle is 10 cm long, and its diagonals each measure 12 cm. Calculate the breadth of the rectangle to one decimal place.

7   A ladder 10 m long has its base 3 m from the foot of a wall, as shown in Figure 9. How far up the wall does the ladder reach? Answer to the nearest centimetre.

Pythagoras' theorem

8   Figure 10 shows the end view of a lean-to shed. Calculate the length of the sloping edge of the roof to the nearest metre.

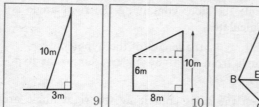
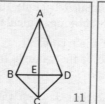
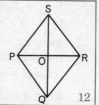

9   In Figure 11, ABCD is a kite. BD is 6 cm long, AC is 12 cm long and AE is 8 cm long. Calculate the lengths of the sides of the kite, to one decimal place if necessary.

10 In Figure 12, PQRS is a rhombus. Each side is 11 cm long and QS is 18 cm long. Calculate the length of PR to one decimal place.

11 A ship sails 6 km north, then 6 km east. How far is it from its starting point to the nearest km?

12 Figure 13 shows a cube with each edge 5 cm long.
    a What is the shape of each face of the cube?
    b Name the right angle in triangle BCG, and calculate the length of BG.
    c Assuming that BGH is a right angle, calculate the length of BH.

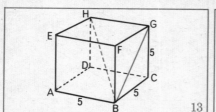
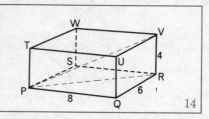

13 Figure 14 shows a cuboid with edges 8 cm, 6 cm and 4 cm long.
    a What is the shape of each face of the cuboid?
    b Name the right angle in triangle PQR, and calculate the length of PR.
    c Assuming that angle PRV = 90°, calculate the length of PV.

Exercise 2B

1   An isosceles triangle has its base 10 cm long and each of its equal sides 13 cm long. Calculate its altitude and its area.

Geometry

2. An isosceles right-angled triangle has its hypotenuse 8 cm long and each of its equal sides x cm long. Find x.

3. In a right-angled triangle the sides containing the right angle are 6 cm and 8 cm long. Calculate:
   a the area of the triangle
   b the length of the hypotenuse
   c the altitude of the triangle from the right-angle vertex to the hypotenuse.

4. Repeat question *3* when the sides containing the right angle are 12 cm and 5 cm respectively.

5. An aircraft flies 120 km south, then 150 km east, then 200 km north. How far is it from its starting point?

6. Figure 15 shows a pyramid on a square base. The edges of the base each measure 10 cm and each slant edge measures 13 cm. Calculate the length of the altitude VM of one of the triangular faces. Calculate, correct to one decimal place, the vertical height VO of the pyramid from V to the centre O of the square.

7. In Figure 15 calculate the length of VO by another method. (*Hint*: First find d^2 where OC = d cm.)

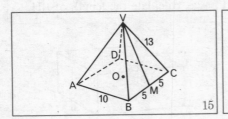
15

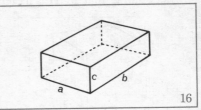
16

8. A pyramid like that in Figure 15 is constructed. The base is a square of side 16 cm and the vertical height is 15 cm. Calculate the length of the altitude of a triangular face. Hence calculate the length of VB to the nearest millimetre.

9. Figure 16 represents a cuboid with edges a, b and c units long as shown. If s units is the length of a space diagonal, find a formula connecting s^2, a^2, b^2 and c^2. (It will help you if you think of the method of solving questions *12* and *13* of Exercise 2.)

3 Using coordinates. The distance formula

We saw in Section 2 that we could calculate the length of the hypotenuse of a right-angled triangle. In Section 1 we saw that we could calculate lengths from the number of units across and the number of units up or down. In this section we put the two results together in the coordinate plane.

Exercise 3

1. In Figure 17(i) the coordinates of A and B are shown. AC and CB are drawn parallel to the axes, and along with AB form a right-angled triangle.
   a. Write down the lengths of AC and CB.
   b. Use Pythagoras' theorem to calculate the length of AB.

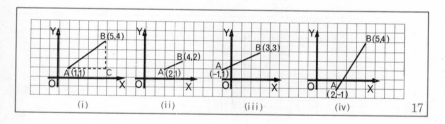

2. Repeat question *1* for Figure 17(ii).
3. Repeat question *1* for Figure 17(iii).
4. Repeat question *1* for Figure 17(iv).

The Distance Formula

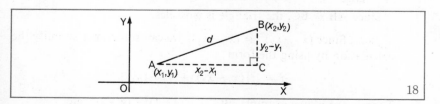

Geometry

In Figure 18, A is the point (x_1, y_1) and B is the point (x_2, y_2).
The length of AC = $x_2 - x_1$.
The length of CB = $y_2 - y_1$.
If d is the length of AB, then, by Pythagoras' theorem,

$$d^2 = (x_2 - x_1)^2 + (y_2 - y_1)^2.$$

Hence $\quad d = \sqrt{[(x_2 - x_1)^2 + (y_2 - y_1)^2]}.$

This is called the *Distance Formula*. It is used to calculate the distance between any two points on the coordinate plane.

Figure 18 shows the easy case where A and B are both in the first quadrant and AB slopes up from left to right. The formula, however, can be used in all cases as the magnitude of AC is always given by $x_2 - x_1$ and that of CB by $y_2 - y_1$. You can verify this by trying examples for yourself on squared paper.

Example. By calculating the lengths of the sides AB and BC of the triangle with vertices A $(-2, 1)$, B $(4, 4)$ and C $(1, -2)$ show that the triangle is isosceles.

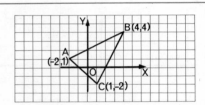

19

$$\begin{aligned}
AB &= \sqrt{(4-(-2))^2 + (4-1)^2} & BC &= \sqrt{(1-4)^2 + (-2-4)^2} \\
&= \sqrt{6^2 + 3^2} & &= \sqrt{(-3)^2 + (-6)^2} \\
&= \sqrt{36 + 9} & &= \sqrt{9 + 36} \\
&= \sqrt{45} & &= \sqrt{45}
\end{aligned}$$

Since AB = BC, the triangle is isosceles.

Note: Since $(x_1 - x_2)^2 = (x_2 - x_1)^2$ we can sometimes simplify the calculation by using the form

$$d = \sqrt{[(x_1 - x_2)^2 + (y_1 - y_2)^2]}.$$

Then in the above example, BC = $\sqrt{(4-1)^2 + (4+2)^2}$, etc.

Using coordinates. The distance formula

Exercise 4

1. Check your answers to the questions in Exercise 3 by using the distance formula to calculate the lengths of the lines joining:
   a. A (1, 1) and B (5, 4)
   b. A (2, 1) and B (4, 2)
   c. A (−1, 1) and B (3, 3)
   d. A (2, −1) and B (5, 4).

2. Calculate the distances between the following pairs of points.
   a. The origin and P (3, 4)
   b. The origin and Q (8, −6)
   c. O (0, 0) and R (−5, −12)
   d. S (−2, 5) and T (3, −7)
   e. U (4, −11) and V (−5, 1)
   f. W (−1, −1) and Z (−4, −4)

3. The vertices of a triangle are A (4, 4), B (−2, 3) and C (3, −2). Use the distance formula to show that ABC is an isosceles triangle.

4. Show that the points (5, 0), (4, 3), (0, 5), (−3, 4), (−5, 0), (−3, −4), (0, −5) and (4, −3) all lie on a circle with its centre at the origin. Illustrate by a sketch.

5. State which of the following expressions represent the distance between the points (p, q) and (r, s).
   a. $\sqrt{(r-p)^2+(s-q)^2}$
   b. $\sqrt{(p-q)^2+(r-s)^2}$
   c. $\sqrt{(p-r)^2+(q-s)^2}$
   d. $\sqrt{(q-s)^2+(p-r)^2}$

6. a. On squared paper plot the points P (−4, −1), Q (−3, −2), R (0, 1) and S (−1, 2).
   b. Calculate the lengths of the sides and diagonals of PQRS.
   c. Use your answers to state the shape of PQRS.

Exercise 4B

1. Calculate the lengths of the four sides and the two diagonals in the quadrilaterals whose vertices are:
   a. A (−1, 1), B (3, 2), C (5, 5) and D (1, 4)
   b. A (1, −2), B (5, 0), C (7, 4) and D (3, 2).

   From the lengths you have calculated, deduce what kind of figure each quadrilateral must be.

2. Show that the points (5, 6), (7, 2), (−1, 6), (−3, 2), (−2, −1) and (6, −1) all lie on a circle with centre (2, 2).

Geometry

3 Calculate the distances between:

  a $(p, 4p)$ and $(-2p, 8p)$ *b* $(-2m, 5m)$ and $(-4m, -2m)$

4 A ship S is situated 4 km east and 2 km north of a port P and another ship T is situated 7 km east and 8 km north of P.

Illustrate on a diagram and calculate the distance between the ships to two significant figures. (*Hint*: Think of coordinates with P as origin and axes pointing east and north.)

5 $(3, 1)$ is the centre of a circle and $(8, 13)$ is a point on its circumference. Calculate the length of the radius of the circle.

Use the distance formula to show that the point $(-9, -4)$ must lie on the circumference of the same circle.

6 $(0, 3)$ is the centre of a circle and $(6, 5)$ is a point on the circumference. Calculate the length of the radius of the circle.

Draw the circle on squared paper and use the square on the radius to calculate the coordinates of the points in which the circle cuts the x-axis.

4 The converse of Pythagoras' theorem

The word *converse* is used in a special way in mathematics. Look at the following examples.

If $3x - 5 = 1$, *then* $x = 2$.
Converse. *If* $x = 2$, *then* $3x - 5 = 1$.
In triangle ABC, *if* AB = AC, *then* angle B = angle C.
Converse. In triangle ABC, *if* angle B = angle C, *then* AB = AC.

To write down a converse all we need to do is to interchange the 'if' part of the sentence and the 'then' part. It follows that in the above pairs of results, each is the converse of the other.

We may write down a converse but it does not have to be true, e.g.
Statement: *If* a number ends in zero, *then* it is divisible by 5. (True)
Converse: *If* a number is divisible by 5, *then* it ends in zero. (False)
Statement: *If* $x = 5$, *then* $x^2 = 25$. (True)
Converse: *If* $x^2 = 25$, *then* $x = 5$. (False if $x \in R$)
(What other replacement is possible for x?)

The converse of Pythagoras' theorem

Exercise 5

Give the converse of each of the following true statements, and say whether each converse is true or false.

1. *If* a closed shape has four angles, *then* it has four sides.
2. *If* x is a prime number greater than 2, *then* x is an odd number.
3. *If* a quadrilateral is a square, *then* its diagonals intersect at right angles.
4. *If* $x = 5$, *then* $2x - 3 = 7$.
5. *If* an angle is acute, *then* the supplement of the angle is obtuse.
6. *If* a boy lives in London, *then* he lives in England.
7. *If* x and y are odd numbers, *then* $x + y$ is even.
8. *If* AB and CD are both perpendicular to XY, *then* AB and CD are parallel.
9. *If* triangle ABC has three equal sides, *then* it has three equal angles.
10. *If* $x \in A \cap B$, *then* $x \in A$.

These examples show that we cannot write down the converse of a true theorem and claim that this converse is true.

Pythagoras' theorem states

'In triangle ABC, *if* angle A is a right angle, *then* $a^2 = b^2 + c^2$.'

The converse states

'In triangle ABC, *if* $a^2 = b^2 + c^2$, *then* angle A is a right angle.'

We know that the theorem is true. Is the converse also true? Look at Figure 20.

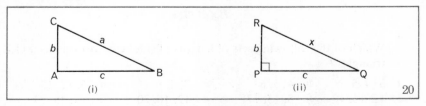

20

In Figure 20(i), $a^2 = b^2 + c^2$. Is angle BAC = 90°?

In Figure 20(ii), PQ = c, PR = b and angle QPR is a right angle.

    In (i) $a^2 = b^2 + c^2$ (given)

    In (ii) $x^2 = b^2 + c^2$ (Pythagoras' theorem)

    Hence $x^2 = a^2$ and $x = a$.

Geometry

The three sides of triangle ABC are equal respectively to the sides of triangle PQR; hence the triangles are congruent. (We saw this in the section on construction of triangles.)

Hence angle BAC = angle QPR = 90°.

Hence in triangle ABC, if $a^2 = b^2 + c^2$, then angle A is a right angle. This means that we can use the converse of Pythagoras' theorem when we know the lengths of the sides of a triangle to check if the triangle is right-angled or not.

Example 1. Show that the triangle with sides of lengths 2·5 cm, 2 cm and 1·5 cm is right-angled.

Let $a = 2\cdot5$, $b = 2$ and $c = 1\cdot5$.

Then $a^2 = 2\cdot5^2 = 6\cdot25$

$\left.\begin{array}{l} b^2 = 2^2 = 4 \\ c^2 = 1\cdot5^2 = 2\cdot25 \end{array}\right\} b^2 + c^2 = 4 + 2\cdot25 = 6\cdot25$

Since $a^2 = b^2 + c^2$, the triangle is right-angled (by the converse of Pythagoras' theorem).

Example 2. The sides of a triangle are 9, 7 and 6 units long. Is the triangle right-angled?

Let $a = 9$, $b = 7$ and $c = 6$.

Then $a^2 = 81$

$\left.\begin{array}{l} b^2 = 49 \\ c^2 = 36 \end{array}\right\} b^2 + c^2 = 49 + 36 = 85 \neq 81$

Since $a^2 \neq b^2 + c^2$, the triangle is not right-angled.

Note that it is advisable to consider the longest side by itself and the two shorter sides together.

Exercise 6

1 Which of the following sets of lengths of sides can form right-angled triangles?

- *a* 3, 4, 5
- *b* 4, 5, 6
- *c* 7, 4, 8
- *d* 5, 12, 13
- *e* 12, 15, 19
- *f* 8, 15, 17
- *g* 12, 16, 20
- *h* 28, 45, 53

2 In Figure 21(i):

- *a* calculate the length of AB
- *b* calculate the length of AC
- *c* use the converse of Pythagoras' theorem to prove that angle BAC is right.

The converse of Pythagoras' theorem

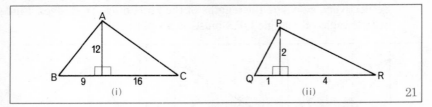

Figure 21

3 As in question 2, prove that in Figure 21(ii) angle QPR is a right angle.

4 In a parallelogram one side measures 13 cm and the diagonals measure 10 cm and 24 cm. Prove that the diagonals are at right angles. What kind of parallelogram must it be?

5 Use the distance formula and the converse of Pythagoras' theorem to show that OA is perpendicular to OB when the coordinates of A and B are respectively:

a (3, 2) and (4, −6) b (4, −2) and (3, 6) c (4, 1) and (−2, 8).

6 A is the point (5, 0), B is (−5, 0) and C is (3, 4). Prove that angle ACB is a right angle.

Exercise 6B

1 Which of the following sets of lengths of sides can form right-angled triangles?

a 2, 3, 4 b 9, 40, 41 c 1, 2, $\sqrt{3}$ d 1, 2, $\sqrt{5}$
e 20, 21, 29 f 0·6, 0·8, 1 g 25, 30, 40 h 23, 264, 265

2 A is the point (−2, −1), B is (2, 2) and C is (5, −2). Prove that angle ABC is a right angle.

3 In triangle XYZ, XU is drawn perpendicular to YZ. If UY = 25, UX = 60 and UZ = 144 units, prove that YXZ is a right angle.

4 Figure 22 shows a tetrahedron ABCD (a solid figure with four triangular faces). ABD, ABC and BDC are right angles. AB, BD and DC measure 9, 12 and 16 units. Calculate the length of each of the other three edges. Prove that angle ADC is a right angle.

5 Figure 23 shows a tetrahedron PQRS in which QPR, RPS and PSQ are right angles. If the edges PQ, PR and PS measure 7, 4 and 3 units respectively, calculate the lengths of the remaining three edges. Find which angle of triangle QRS must be a right angle.

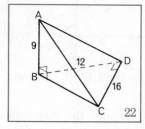

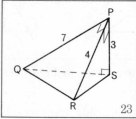

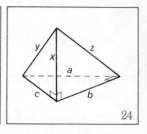

6   Figure 24 shows a tetrahedron with three of its faces right-angled as marked on the diagram; lengths of edges are also marked on the diagram. Use Pythagoras' theorem to find a relation between y, z and a. Find two other edges that are perpendicular to each other.

Summary

1 *Pythagoras' theorem*

The square on the hypotenuse of a right-angled triangle is equal to the sum of the squares on the other two sides.

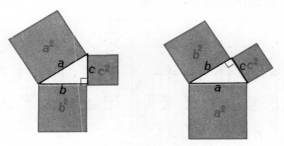

OR In triangle ABC, *if* angle A is a right angle, *then* $a^2 = b^2 + c^2$

2 *The Distance Formula*

$$d = \sqrt{[(x_2 - x_1)^2 + (y_2 - y_1)^2]}$$

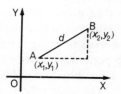

3 *Converse of Pythagoras' theorem*

In triangle ABC, *if* $a^2 = b^2 + c^2$, *then* angle A is a right angle.

Geometry

Topics to explore

(i) Strip patterns

Each line of Figure 1 is to be thought of as forming part of an endless frieze which extends indefinitely in both directions, giving an infinite strip pattern.

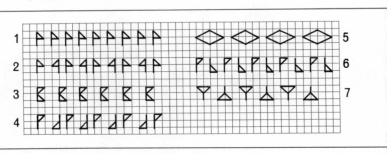

1

If we could make a tracing of such a frieze, we could bring it into coincidence with the frieze by simply sliding it along a suitable distance in the direction of the frieze. (Is there more than one suitable distance?) The pattern is conserved by suitable *translations*—which, notice, leave no points invariant.

Most of the patterns have one or more symmetries in addition. For each pattern, try to discuss in how many different kinds of ways you could transform it so as to bring it into coincidence with itself.

Notice that the sixth pattern is conserved by translation but not by reflection; and also by a combination of both—a new transformation called a *glide reflection*. The glide reflection occurs again in the last line, along with reflections in certain *axes* and also in certain *points*.

You may be interested to know that Figure 1 illustrates all the possible *kinds* of frieze patterns that can occur—there are only seven. (Of course, the detail or 'motif' of the pattern is infinitely variable, but the pattern can repeat itself, along the line of the frieze, in only seven recognizably different ways.) Try to find examples of all seven types in wallpaper friezes, in embroidery, or in architecture, pottery, etc.

Note that the lines of letters ... S S S S S S ... and p d p d p d ... are patterns of the same form as line four; and the row ... p b p b p b .. is just like line 6. Make up other such rows for these and other lines.

Topics to explore

(ii) Pythagorean triples

In the examples you have tried in this chapter you found that in some cases it was possible to get an exact whole number for the number of units in the length of the hypotenuse as well as for the sides containing the right angle. We had, for example, right-angled triangles with sides 3, 4 and 5 units, or 5, 12 and 13 units. We found that multiples of these like 9, 12, 15 or 10, 24, 26 would do equally well. Such *Pythagorean triples* have interested mathematicians since the time of Pythagoras. Indeed long before that the 3, 4, 5 triangle was used to measure right angles. You may experiment with the following table by continuing it as far as you like. Be sure to keep $m > n$.

m	n	m^2+n^2	m^2-n^2	$2mn$	Triple
2	1	$4+1 = 5$	$4-1 = 3$	$2 \times 2 \times 1 = 4$	5, 3, 4
3	1	$9+1 = 10$	$9-1 = 8$		
3	2	$9+4 = 13$			
4	1				
4	2				
4	3				
5	1				
⋮	⋮				

Does each set of answers provide lengths of sides for a right-angled triangle?
Does the same shape of triangle occur more than once?
Do new shapes continue to appear?

Revision Exercises

Revision Exercise on Chapter 1
Translation

Revision Exercise 1

1. Mark a point O near the centre of a page, and draw a line ON to indicate the direction due north. With a scale of 1 cm to represent 1 km, draw lines representing OP, OQ and OR, given that:
   - a. P is 5 km from O on a bearing of 028°
   - b. Q is 4 km from O on a bearing of 175°
   - c. R is 7 km from O on a bearing of 305°.

2. O is a fixed landmark. A is 8 km east and 6 km north of O. B is 15 km east and 8 km south of O. C is 5 km west and 12 km north of O.
   - a. Make a scale drawing on squared paper to show O, A, B and C.
   - b. Measure the distances and three-figure bearings of A, B and C from O.

3. An aircraft is flying straight to its destination on a bearing 043°. If the aircraft is above cloud, how will the pilot know when he has reached his destination?

4. An aircraft flies from an airport A in a direction 042° for 80 km to B. From B it flies on a bearing of 160° for 100 km to C.
   Make a scale drawing of the whole journey, and measure the distance and bearing of C from A.

5. Q is 10 km east and 5 km north of P. R is 2 km west and 10 km north of P.
   - a. Make a scale drawing on squared paper, and measure the distance and bearing of R from Q.
   - b. How far east, and how far south, is Q from R?

Revision Exercise on Chapter 1

6 Complete the following for the diagram in Figure 1.

    a $\vec{BA} \oplus \vec{AC} = ...$ b $\vec{BA} \oplus \vec{AD} \oplus \vec{DC} = ...$

    c $\vec{AB} \oplus \vec{BD} = ... = \vec{AC} \oplus ...$ d $\vec{CA} \oplus \vec{AB} \oplus \vec{BD} \oplus \vec{DA} = ...$

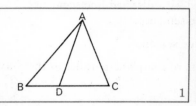
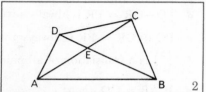

7 Complete the following for the diagram in Figure 2.

    a $\vec{DE} \oplus \vec{EB} = ...$ b $\vec{DE} \oplus \vec{EC} = ...$ c $\vec{DE} \oplus ... = \vec{DA}$

    d $... \oplus \vec{EA} = \vec{CA}$ e $\vec{DE} \oplus \vec{EC} + ... = \vec{DB}$

    f $\vec{AE} \oplus ... \oplus \vec{BC} = \vec{AC}$

8 Copy and complete:

    a In triangle ABC, $\vec{AB} \oplus \vec{BC} \oplus \vec{CA} = ...$

    b In quadrilateral ABCD, $\vec{AB} \oplus \vec{BC} \oplus \vec{CD} \oplus \vec{DA} = ...$
Can you state a result about directed line segments which would be true for every polygon?

9 Plot the points A (3, 2) and B (6, 4) on squared paper. Draw, and mark with an arrow, several directed line segments equal to $\vec{AB}$, and several equal to $\vec{BA}$.

10 Draw a parallelogram. Illustrate by successive translations how the plane can be tiled completely by congruent parallelograms.

11a On squared paper plot the points A (4, 1), B (8, 3), C (10, 8) and D (6, 6). What kind of figure is ABCD?

    b The figure is given a translation $\begin{pmatrix} 4 \\ -1 \end{pmatrix}$. Write down the coordinates of the images of A, B, C and D, and show the new figure in your diagram.

    c Another translation maps C to the origin O. Write down the coordinates of the images of A, B and D.

Geometry

12 State whether each of the following is true or false.
  - **a** A bearing of north-west can be given by 045°.
  - **b** A directed line segment is described completely if its magnitude (or length), direction and initial point are given.
  - **c** A translation can be represented by a directed line segment.
  - **d** PQ+QR = PR is always a true sentence.
  - **e** $\vec{PQ} \oplus \vec{QR} = \vec{PR}$ is always a true sentence.
  - **f** $\begin{pmatrix}5\\0\end{pmatrix}$ is a translation of all points of the plane parallel to the x-axis.
  - **g** If $\vec{AB} = \vec{CD}$, then:
    - (*1*) AB = CD
    - (*2*) AB ∥ CD
    - (*3*) $\vec{AC} = \vec{DB}$
    - (*4*) AD = BC
    - (*5*) ABDC is a parallelogram.

13a Draw a circle with centre A (1, 2) and radius 4 units.
  - **b** Draw the image of the circle under the translation $\begin{pmatrix}-1\\-2\end{pmatrix}$.
  - **c** Mark four points on the original circle, and show their images on the image circle.

14a Draw the shapes with vertices A (−12, 2), B (−11, 9), C (−8, 3), D (−9, −4) and P (−4, −2), Q (−3, 5), R (0, −1), S (−1, −8).
  - **b** State the components of the translations which would map:
    (*1*) ABCD to PQRS (*2*) PQRS to ABCD.

15 F is the point (3, 2), G is (5, 6) and M is the midpoint of FG.
  - **a** Give the components of the translations represented by $\vec{FG}$ and $\vec{FM}$.
  - **b** Give the coordinates of M and of its image under the translation $\begin{pmatrix}8\\8\end{pmatrix}$.
  - **c** Describe the position of M with respect to F and G under any translation.

16 Draw a circle with centre (4, 0) and radius 6 units. Under what translations will the circle pass through O and still have its centre on the x-axis? Draw the image in each case.

17 A is the point (−2, −4), and B is (8, −4), P = {(3, 8), (3, −10)}.
  - **a** What shape of triangle is formed with base AB and either member of set *P* as third vertex?
  - **b** Write down the components of the translation which maps the members of *P* to the y-axis.

Revision Exercise on Chapter 1

c Write down the components of two translations, each of which maps one member of *P* to the *x*-axis. Give the coordinates of the images of A and B in each case.

18 A is the point (7, −3), and B is (4, 1).

a Write down the components of the translation which maps A to the origin.
b Write down the coordinates of the image of B under this translation.
c Another translation maps B to the origin. Give its components, and also the coordinates of the image of A under this translation.

19 With the aid of Figure 3, draw directed line segments on squared paper to represent the following:

a $\vec{AB} \oplus \vec{CD}$ *b* $\vec{AB} \oplus \vec{EF}$ *c* $\vec{AB} \oplus \vec{GH}$
d $\vec{CD} \oplus \vec{EF}$ *e* $\vec{CD} \oplus \vec{GH}$ *f* $\vec{EF} \oplus \vec{GH}$

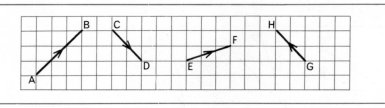

3

20 Calculate the components of the single translations resulting from:

a $\begin{pmatrix} -1 \\ -2 \end{pmatrix} + \begin{pmatrix} 2 \\ -3 \end{pmatrix}$ *b* $\begin{pmatrix} -1 \\ -2 \end{pmatrix} + \begin{pmatrix} 3 \\ -1 \end{pmatrix}$ *c* $\begin{pmatrix} 2 \\ -3 \end{pmatrix} + \begin{pmatrix} -3 \\ -1 \end{pmatrix}$

d $\begin{pmatrix} 4 \\ 3 \end{pmatrix} + \begin{pmatrix} -1 \\ 2 \end{pmatrix} + \begin{pmatrix} -2 \\ -3 \end{pmatrix}$ *e* $\begin{pmatrix} 1 \\ 0 \end{pmatrix} + \begin{pmatrix} 0 \\ 3 \end{pmatrix} + \begin{pmatrix} -2 \\ -2 \end{pmatrix} + \begin{pmatrix} 1 \\ -1 \end{pmatrix}$

21*a* Calculate the components of the single translations given by:

(*1*) $\begin{pmatrix} -2 \\ -3 \end{pmatrix} + \begin{pmatrix} 2 \\ 3 \end{pmatrix}$ (*2*) $\begin{pmatrix} p \\ q \end{pmatrix} + \begin{pmatrix} -p \\ -q \end{pmatrix}$ (*3*) $\begin{pmatrix} a-b \\ c+d \end{pmatrix} + \begin{pmatrix} b-a \\ -c-d \end{pmatrix}$

b Why could we describe $\begin{pmatrix} -a \\ -b \end{pmatrix}$ as the negative of $\begin{pmatrix} a \\ b \end{pmatrix}$?

22*a* If $\vec{AB}$ represents $\begin{pmatrix} 3 \\ -2 \end{pmatrix}$ and $\vec{CD}$ represents $\begin{pmatrix} -1 \\ 5 \end{pmatrix}$, state the components of the translations represented by:

(*1*) $\vec{AB} \oplus \vec{CD}$ (*2*) $\vec{AB} \oplus \vec{DC}$ (*3*) $\vec{BA} \oplus \vec{CD}$ (*4*) $\vec{BA} \oplus \vec{DC}$

b Can you see a connection between pairs of the answers?

Geometry

Revision Exercises on Chapter 2
The Calculation of Distance

Revision Exercise 2A

1. Using your square root tables if necessary, calculate the lengths of sides of squares which have the following areas in square units.

   a 25   b 81   c 121   d 30   e 65   f 180

2. On squared paper draw a straight line AB which goes 6 to the right and 3 up. Draw the square on AB and the 'surrounding' square. Calculate the area of the square on AB and the length of AB.

3. Use Pythagoras' theorem to calculate the value of x in each triangle in Figure 4.

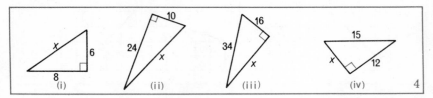
Figure 4

4. In each triangle in Figure 5, calculate x to 3 significant figures.

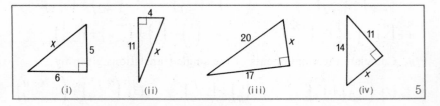
Figure 5

5. Calculate the length of the diagonal of a rectangle which is 9 cm long and 8 cm broad.

6. The diagonals of a rhombus are 16 cm and 12 cm long respectively. Calculate the length of a side of the rhombus.

7. In Figure 6 ABCDEFGH is a cuboid with AB = 12 cm, BC = 10 cm and CG = 8 cm.

Revision Exercises on Chapter 2

    a Calculate the lengths of its face diagonals.
    b What is the shape of the figure AEGC?
    c Calculate the length of the space diagonal AG.

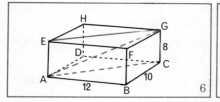

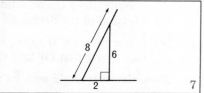

8   A ladder 8 metres long leans against a wall 6 metres high as shown in Figure 7. If the foot of the ladder is 2 metres from the wall, calculate what length projects over the top of the wall.

9   Figure 8 shows the gable end of a house with dimensions in metres as marked on the diagram. If the gable end is symmetrical, calculate the length of AB.

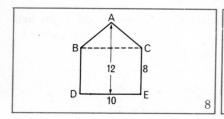

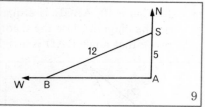

10   In Figure 9, A and B are coastguard observation posts, B being due west of A. From A a ship S is observed due north and 5 km distant. From B the ship is 12 km distant. Calculate the distance between A and B.

11   A is the point (1, 2), B is (4, 6), C is (−1, 2) and D is (−5, −1).
    a Use the distance formula to show that AB = DC.
    b Which is longer, BC or AD?

12   A is the point (2, 1), B is (3, 8), C is (−2, −1) and D is (−7, 4).
    a Show that AB = DC.
    b Calculate the difference in length between AD and BC.

13   (2, 4) is the centre of a circle and (−1, −3) lies on the circumference. Calculate the length of the radius of the circle.

Use the distance formula to determine if the point $(-2, -2)$ lies inside, on, or outside the circumference of the circle.

14 In a rectangle ABCD, A is the point $(-1, 2)$, B is $(1, -2)$ and C is $(7, 1)$. Write down the coordinates of D and calculate the lengths of the sides and of the diagonals of the rectangle.

15 If P is the point $(3, 4)$ and Q is $(8, -6)$, calculate the lengths of OP and OQ. Prove that OP and OQ are perpendicular.

16 Show that a triangle with its sides 5, 12 and 13 units long is right-angled.

Without further working state which of the following can be the number of units in the lengths of sides of right-angled triangles.

    a 15, 36, 39   *b* 25, 60, 65   *c* $5p, 12p, 13p$

17 ABCD is a parallelogram in which A is the point $(-1, 4)$, B is $(6, 5)$ and the diagonals intersect at E $(0, 2)$. Illustrate by a diagram on squared paper, and use the converse of Pythagoras' theorem to prove that the parallelogram is a rhombus.

18 In Figure 10, ABCD is a quadrilateral with dimensions as marked on the diagram. Use the theorem of Pythagoras and its converse to prove that angle CAD is a right angle.

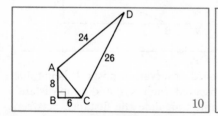
10

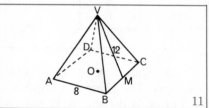
11

19 Figure 11 shows a pyramid on a square horizontal base. The vertex V is directly above the centre O of ABCD. VM is an altitude of triangle VBC and is 12 cm long. Each side of the square base is 8 cm long. Calculate the vertical height of the pyramid and the length of VB.

Revision Exercise 2B

1 A ship S lies 6 km from a port P on a bearing 023°. Another port Q is 15 km distant from P on a bearing 293°. Calculate the distance from S to Q to 3 significant figures.

Revision Exercises on Chapter 2

2. The two *longer* sides of a right-angled triangle measure 60 and 61 units. Calculate the length of the shortest side.
   Calculate also the area of the triangle and the length of the altitude perpendicular to the hypotenuse.

3. A man swimming steadily across a river 60 metres wide is carried 40 metres downstream by a uniform current. Calculate the distance he travels to the nearest metre.

4. A rectangular field measures 203 m by 115 m. Calculate to the nearest metre what distance a man would save by walking from corner to corner along a diagonal instead of going along two sides.

5. Figure 12 shows the cross-section of a swimming pool, 25 m long, 1 m deep at the shallow end, and 3 m deep at the deep end. The floor, s metres long, slopes uniformly from one end to the other. Calculate s, and show that the length of the sloping bottom differs from the horizontal length by only a few centimetres.

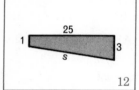
12

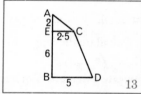
13

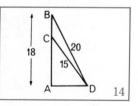
14

6. Figure 13 shows a vertical pole BA standing on horizontal ground BD, supported by a stay wire ACD which goes round the end of a horizontal bar EC. The lengths of AE, EB, EC, BD are 2, 6, 2·5 and 5 metres respectively. Calculate the total length of the stay wire to 2 significant figures.

7. Figure 14 shows a vertical pole AB, 18 m tall, standing on horizontal ground AD. BD and CD are stay wires 20 m and 15 m long. Calculate the lengths of AD, AC and BC to one decimal place.

8. An aeroplane flies 10 km east and 10 km north. Calculate its distance from its starting point.
   Repeat when the aeroplane flies:

   a 100 km in each direction   *b* 1000 km in each direction.
   Why must your answers become increasingly inaccurate?

Geometry

9. Construct accurately a right-angled triangle with the sides containing the right angle each 10 cm long. Measure the hypotenuse. From your answer deduce an approximation for $\sqrt{2}$.

   Make constructions to find the following square roots approximately by measurement: *a* $\sqrt{5}$ *b* $\sqrt{3}$

10. A cyclist has the choice of cycling along two good roads at right angles, 10 km and 15 km long respectively, or along the hypotenuse by a rough road. If his average speed on the good roads is 24 km/h and on the rough road 18 km/h, which is the quicker way to his destination?

11. In a rhombus ABCD, A is the point $(-3, -3)$, B is $(0, 2)$ and D is $(2, 0)$. Find the coordinates of C.

    Calculate the lengths of the diagonals and the area of the rhombus, each to one decimal place.

12. Calculate the distances between the pairs of points:

    a $(5p, -3p)$ and $(0, 9p)$ *b* $(4m, -3m)$ and $(-4m, 12m)$.

13. A translation $\begin{pmatrix} 5 \\ 12 \end{pmatrix}$ is applied to all points in the coordinate plane. A' is the image of A. Calculate the length of AA' when A is the point

    a $(-2, 3)$ *b* $(5, -4)$ *c* (p, q).
    Do you need a separate calculation in each case?

14. Which of the following sets of lengths of sides can form right-angled triangles?

    a $\sqrt{2}, \sqrt{2}, 2$ *b* 5, 7, 8   *c* 6, 3, $3\sqrt{3}$
    d 6, 7, 9   *e* 3333, 4444, 5555

15. Show that the points A $(-3, -2)$ and B $(4, -3)$ lie on the circumference of a circle with centre C $(1, 1)$.

    Write down the coordinates of M, the midpoint of AB. Prove that CM is perpendicular to AB.

16. A pyramid stands on a base which is a regular hexagon of side 10 cm, and the vertex is directly above the centre of the base.

    a If the hexagon is regarded as made up of six equilateral triangles, calculate the altitude of each triangle.

    b If the vertical height is 20 cm, calculate the length of a slant edge and the altitude of one of the slanting triangular faces.

Revision Exercises on Chapter 2

17 Figure 15 shows a triangular pyramid which has been obtained by sawing off a corner of a wooden cuboid. A is a vertex of the original cuboid. AX, AY and AZ measure 5, 5 and 7 units respectively. Calculate the lengths of XY, YZ and ZX to two significant figures.

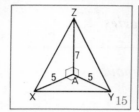

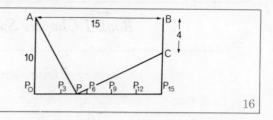

18 Figure 16 represents a rectangular courtyard measuring 15 m by 10 m. A rope is to be stretched from A to some point P on the wall P_0P_{15} and from there to C, 4 m from B. We wish to find the shortest possible rope.

Make a table of the lengths of AP+PC for the positions $P_0, P_3, \ldots, P_{15}$ which are equally spaced along the 15 m wall.

Draw a graph of the length AP+PC against the distance of P from P_0, and read off the shortest length of AP+PC from your graph.

Cumulative Revision Section (Books 1-4)

Book 1 Chapter Summaries

Chapter 1 Cube and cuboid

1 Each face of a *cuboid* is a *rectangle*

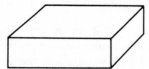

2 Each face of a *cube* is a *square*.

3 Cubes and cuboids each have 3 sets of *parallel* edges. They can be constructed from suitable *nets*.

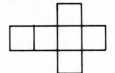

4 Tiles of the same shape and size are *congruent* to each other.

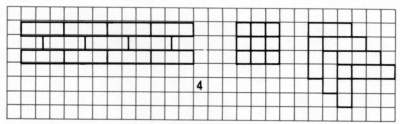

5 Some shapes have *half turn symmetry*.

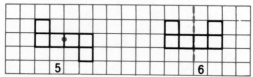

6 Some shapes have *line symmetry*.

Chapter 2 Angles

1 Definitions

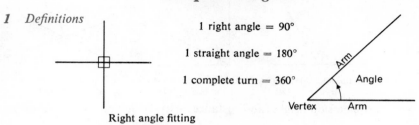

1 right angle = 90°

1 straight angle = 180°

1 complete turn = 360°

Right angle fitting property

2 Kinds of angles

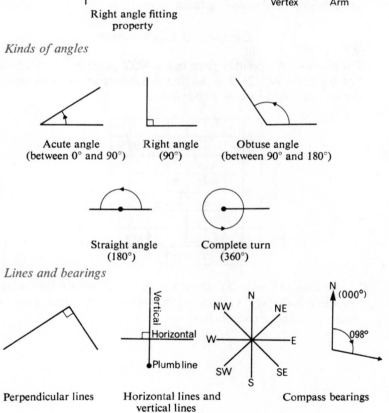

Acute angle (between 0° and 90°)   Right angle (90°)   Obtuse angle (between 90° and 180°)

Straight angle (180°)   Complete turn (360°)

3 Lines and bearings

Perpendicular lines   Horizontal lines and vertical lines   Compass bearings

Geometry

4 *Related angles*

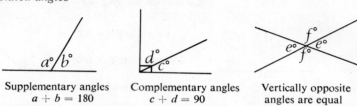

Supplementary angles Complementary angles Vertically opposite
$a + b = 180$ $c + d = 90$ angles are equal

5 Problems on heights and distances may be solved by *scale drawing*.

Chapter 3 Coordinates

The position of a point is given in the XOY plane by its coordinates. For the point A(3, 2), 3 is the first coordinate or *x*-coordinate, 2 is the second coordinate or *y*-coordinate.

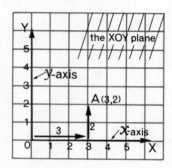

The axes OX and OY are often called Cartesian axes after the seventeenth-century French mathematician Descartes who introduced the idea of coordinates.

Book 2 Chapter Summaries

Chapter 1 Rectangle and square

1 *The rectangle*

Two axioms

1. A rectangle fits its outline in the four ways shown

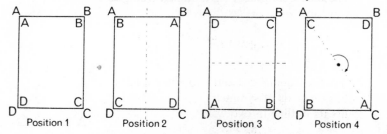

2. Congruent rectangles can be fitted together to cover the plane exactly.

Deductions

a The opposite sides are equal and parallel.
b The diagonals are equal, and bisect each other.
c All the angles are right angles.

2 *The square*

The square is a special kind of rectangle, and fits its outline in eight ways.

Deductions

The square has all the properties of a rectangle, and in addition:

a All the sides are equal.
b The diagonals bisect the angles, and bisect each other at right angles.

3 *Parallel lines*

Parallel lines are straight lines which keep the same distance apart, and so never meet.

Geometry

Chapter 2 Triangles

1. *A rectangle* can be divided into two congruent right-angled triangles.

2. These two congruent right-angled triangles can form *an isosceles triangle.*

3. *An isosceles triangle:*
has two equal sides and two equal angles;
has one axis of symmetry;
fits its outline in two ways.

4. *The sum of the angles of a triangle is 180°,*
i.e. $\angle A + \angle B + \angle C = 180°$.

5. *An equilateral triangle* fits its outline in six ways. It has three equal sides, three equal angles (each 60°), three axes of symmetry.

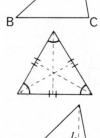

6. *The area of a triangle* is given by:
$\frac{1}{2}$ base × height, i.e. $\triangle = \frac{1}{2}bh$.

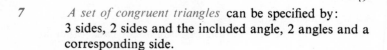

7. *A set of congruent triangles* can be specified by:
3 sides, 2 sides and the included angle, 2 angles and a corresponding side.

Book 3 Chapter Summaries
Chapter 1 Reflection

1. If a figure is mapped onto itself under *reflection in a line*, the line is called an *axis of symmetry*, and the figure has *bilateral symmetry* about the axis.

2. If a figure is mapped onto itself under *a half turn about a point*, the point is called the *centre of symmetry*.
   In both of the above, a figure and its *image* are congruent.

3. Under reflection in XY,
   $A \leftrightarrow A'$ AA' is parallel to BB'
   $B \leftrightarrow B'$ AA' and BB' are bisected at right angles
   $AB \leftrightarrow A'B'$ by XY.
   $AB = A'B'$ AB and A'B' meet on XY, or are
   $\angle ACY = \angle A'CY$ parallel to XY.

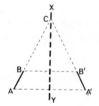
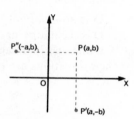

4. Under reflection in the x-axis, $P(a, b) \leftrightarrow P'(a, -b)$
   Under reflection in the y-axis, $P(a, b) \leftrightarrow P''(-a, b)$

5. A *rhombus* consists of two congruent isosceles triangles base to base, and fits its outline in four ways. Also
   the diagonals are axes of symmetry;
   all the sides are equal;
   the opposite angles are equal, and are bisected by the diagonals;
   the diagonals bisect each other at right angles.

Geometry

6   A *kite* consists of two isosceles triangles with equal bases, and fits its outline in two ways. Also
   one diagonal is an axis of symmetry;
   two pairs of sides are equal;
   one pair of opposite angles is equal;
   one diagonal bisects the other at right angles.

7   *Lines constructed* in a triangle:

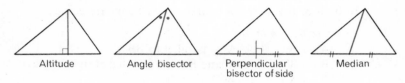

Altitude   Angle bisector   Perpendicular bisector of side   Median

Chapter 2 The parallelogram

1   Under a *half turn* about O (or *reflection* in O):
   O is an invariant point;
   O is the midpoint of the line joining a point and its image;
   a line and its image are parallel, or in the same line, but their directions are opposite.

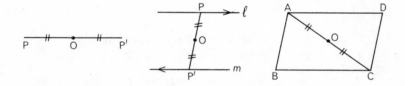

2   A *parallelogram* is formed by a triangle and its image under a half turn about the midpoint of one side. Also:
   the opposite sides are equal and parallel;
   the opposite angles are equal;
   the diagonals bisect each other;
   it is conserved under a half turn about the point of intersection of the diagonals;
   it can be formed by drawing two pairs of parallel lines.

3   A plane can be covered by a tiling *of congruent parallelograms* containing a *tiling* of congruent triangles.

Book 3 Chapter Summaries

(i)   (ii)

4 If a straight line cuts two parallel lines:
   (i) *corresponding angles* are equal;
   (ii) *alternate angles* are equal.

5 *The area of a parallelogram* = base × perpendicular height.

Chapter 3 Locus, and equations of a straight line

1 A *locus* can be regarded as a set of points defined in some way, an important special case being the path traced out by a moving object.

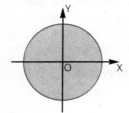
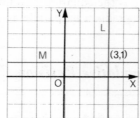
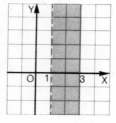

{P: OP ⩽ 1 cm} $L = \{(x, y): x = 3\}$, $\{(x, y): 1 < x \leq 3\}$
                  $M = \{(x, y): y = 1\}$
                  $L \cap M = \{(3, 1)\}$

2 *Equations of a straight line*

(i) $y = x$ is the equation of a straight line through the origin, at 45° to OX.

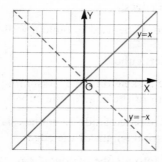
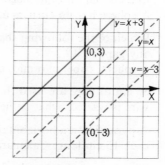

(ii) $y = x + c$ is the equation of a line parallel to $y = x$, through the point $(0, c)$.

Geometry

(iii) $y = mx$ is the equation of a line through the origin, with gradient m.

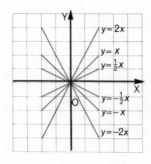 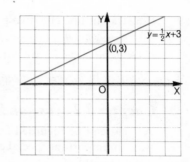

(iv) $y = mx + c$ is the equation of a line through the point $(0, c)$, with gradient m, parallel to the line $y = mx$.

Cumulative Revision Exercises

Exercise A

1   State whether each of the following is true or false.
   a Every set of six congruent squares can be formed into the shape of a cube.
   b In a pyramid on a square base the number of corners + the number of faces − the number of edges = 2.
   c In a cuboid all four space diagonals are equal.
   d A sphere has no corners and no edges.

2   Which of the following shapes could be used to build a wall without any gaps in it?
       sphere, cube, pyramid, cylinder, cuboid

3   36 cm of wire are used to make a skeleton cube. Assuming that no wire is wasted, what is the length of the edge of the cube?

4   A pyramid on a horizontal square base of side 12 cm has slant edges each 15 cm long. Calculate the total length of the edges.

5   Sketch a net that can be folded into a cube of edge 2 cm.

Cumulative Revision Exercises

6 Sketch a net that can be folded into a cuboid which is 3 cm by 2 cm by 1 cm.

7 Write down the set of capital letters in the English alphabet which have both half turn symmetry and line symmetry.

8 Draw a quadrilateral of any shape and measure its angles. (Make your drawing fairly large. Why?)
What is the sum of the angles of the quadrilateral?

9 What is the smaller angle between the hands of a clock at:
a 3 o'clock   *b* 6 o'clock   *c* 8 o'clock   *d* half past 5?

10 In Figure 1, name *a* an acute angle   *b* an obtuse angle
c a pair of vertically opposite angles *d* a pair of supplementary angles.

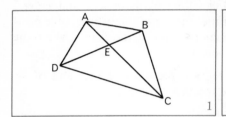
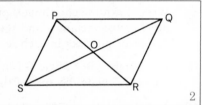

11 In Figure 2, PQRS is a parallelogram. Name eight pairs of supplementary angles in the figure.

12 State which of the following pairs of angles are complementary, supplementary, or neither.
a 32° and 148°   *b* 42° and 148°   *c* 36° and 54°   *d* 120° and 240°
e $p°$ and $(180-p)°$, where $p < 180$ *f* $x°$ and $3x°$, where $x = 22\frac{1}{2}$

13 Which of the following statements are true and which are false?
a SW is the same as a bearing of 225°.
b A ship sailing NE has the same bearing as a NE wind.
c If a ship sailing SE turns through an anticlockwise angle of 30°, its new bearing is 165°.
d An aircraft flying from W to E is on a bearing of 090°.
e The least necessary change of direction from a bearing of 020° to a bearing of 340° is 320°.

14 Draw a triangle PQR where PQ = 5·5 cm, angle RPQ = 50° and angle PQR = 70°. Measure QR and RP.

Geometry

15 Which of the following statements are true and which are false?

   a Since a square fits into its outline in eight ways, it must have eight axes of symmetry.

   b A rectangle is the only kind of quadrilateral which fits into its own outline in exactly four ways.

   c If S is the set of squares and R the set of rectangles, then $S \subset R$.

   d Every face of a cuboid is a rectangle.

16a Draw a rectangle PQRS and show its axes of symmetry.

   b If $\begin{smallmatrix}PQ\\SR\end{smallmatrix}$ shows the order of letters at the start, write down the order of letters after the rectangle is turned over about each axis of symmetry.

   c Write down the order of letters after a half turn about the centre.

17 A rectangle which is 6 units long and 4 units broad has one vertex at the origin and two of its sides along the *positive* x and y-axes. Write down the coordinates of the vertices of the two possible positions of the rectangle and the coordinates of the points where the diagonals intersect in each rectangle.

18 In triangle ABC, angle B is a right angle, AB = 2·5 cm and BC = 6 cm. Calculate the area of the triangle.

19 In Figure 3, the dimensions of the triangles are shown in centimetres.

   a Calculate the area of (i) triangle ABC (ii) triangle PQR.

   b State in each case which of the given measurements are unnecessary for finding the area.

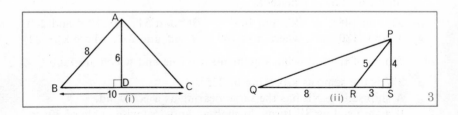

20 PQR is an isosceles triangle in which PQ = PR. If angle PQR = 65°, calculate the size of angle QPR.

21 In Figure 4, copy the diagrams and draw the images of AB, triangle LMN, PQ and square EFGH under reflection in XY in each case. (It will help if you draw your diagrams on squared paper.)

Cumulative Revision Exercises

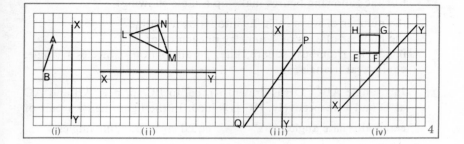

22 The points A (4, 4) and B (4, 2) are mapped to the points A_1 and B_1 under reflection in the x-axis. A and B are also mapped to A_2 and B_2 under reflection in the y-axis. Write down the coordinates of the four images.

  If O is the origin, calculate the areas of triangles OAB, OA_1B_1 and OA_2B_2.

23 P is the point $(-3, 2)$. Write down the coordinates of the images of P under reflection in:

  a the x-axis   *b* the origin   *c* the line $x = 1$ *d* the line $y = x$.

24 Which of the following statements are true and which are false?

  a Every rhombus has two perpendicular diagonals.
  b Every rhombus has two equal diagonals.
  c In every parallelogram the two diagonals bisect each other.
  d Every parallelogram has exactly two axes of symmetry.
  e Each diagonal of a kite bisects the other diagonal.
  f In every kite one diagonal is the perpendicular bisector of the other.

25 Plot the following points on squared paper: A (0, 9), B $(-3, 6)$ and C $(0, -1)$.

  a Find the coordinates of the fourth vertex D of the parallelogram ABCD.
  b Find the coordinates of the fourth vertex D of the parallelogram ABDC.

26*a* Construct a parallelogram PQRS with PQ = 7 cm, QR = 5 cm and angle PQR = 65°. Measure the lengths of the diagonals of PQRS.

  b Make any necessary measurement and calculate the area of PQRS.

Geometry

27 For the following figures make up a table showing:
  a the number of ways each fits into its own outline
  b the number of axes of symmetry in each case;
  rectangle, rhombus, square, kite, parallelogram, isosceles triangle, equilateral triangle.

28 In triangle ABC, DE is drawn parallel to BC to meet AB and AC at D and E respectively. DC is joined. In the figure name a pair of
  a equal alternate angles   b equal corresponding angles.

29 In Figure 5, LM is parallel to QR and the sizes of two angles are given. Copy the diagram and mark in the sizes of all the remaining angles.

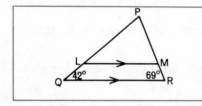
5

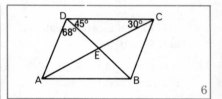
6

30 In Figure 6, ABCD is a parallelogram and the sizes of three angles are given. Copy the diagram and mark in the sizes of all the remaining angles.

31a P is a fixed point 5 cm from a fixed straight line XY. Show on a diagram the set A of points 3 cm from P, and the set B of points 3 cm from XY.
  b Indicate clearly the set $A \cap B$.
  c Show by shading in your diagram the set of points less than 3 cm from P and more than 3 cm from XY.

32 Show by different types of shading on a diagram the sets of points, $x, y \in R$, $A = \{(x, y): 0 \leqslant x \leqslant 4\}$ and $B = \{(x, y): 0 \leqslant y \leqslant 3\}$.
  What shape is the region given by $A \cap B$?

33 On squared paper show the sets of points, $x, y \in R$,
  a $\{(x, y): y = 2x\}$ b $\{(x, y): y = 2x+3\}$ c $\{(x, y): y = 2x-1\}$

34 Suppose the pattern shown in Figure 7 is extended in all directions in the plane. Which of the following translations would conserve the pattern?

a $\vec{AB}$ *b* $\vec{BD}$ *c* $\vec{BH}$ *d* $\vec{GD}$ *e* $\vec{AH}$

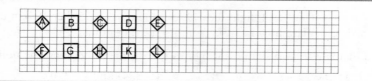

35 Make a scale drawing to represent a displacement $\vec{AB}$ of magnitude 40 km on a bearing 035°, followed by a displacement $\vec{BC}$ of magnitude 45 km on a bearing 300°. Find by measurement the magnitude and bearing of the displacement $\vec{AC}$. (Answer to the nearest km and the nearest degree.)

36 Under a translation, the point A (4, 8) is mapped to the point (10, −2). What is the image of B (8, −2) under the same translation?

37 In triangle ABC, A is the point (4, 7), B is (1, 3) and C is (6, 2). Illustrate on a diagram. Draw the image $A_1B_1C_1$ of the triangle after a translation 3 to the right and 2 up.

38 Under a translation represented by $\vec{OA}$, the points O (0, 0), A (6, 6) and B (6, 3) are mapped to the points O_1, A_1 and B_1 respectively. Find the coordinates of these points and the area of triangle $O_1A_1B_1$.

39 A is the point (3, 2), B is (1, 2) and C is (4, 0). A translation T maps A to the origin O. Under the translation T, C is mapped to C_1. Find the coordinates of C_1. What is the shape of the figure $OBCC_1$?

40 Calculate the magnitudes of the following translations.

a $\begin{pmatrix} 3 \\ -4 \end{pmatrix}$ *b* $\begin{pmatrix} -12 \\ 5 \end{pmatrix}$ *c* $\begin{pmatrix} -2 \\ -2 \end{pmatrix}$ *d* $\begin{pmatrix} p \\ q \end{pmatrix}$

41 Calculate the lengths of the hypotenuses of the following right-angled triangles if the sides containing the right angles have lengths (in cm):

a 12 and 9   *b* 10 and 24   *c* 20 and 21   *d* 7 and 9.

42 Calculate the length of the remaining side in each of the following

Geometry

right-angled triangles where the hypotenuse and one side have lengths (in cm)

   a   17 and 15   *b*   41 and 9   *c*   12 and 7.

43   An aircraft in level flight flies 72 km due north and then 21 km due east. Calculate how far it is from its starting point.

44   Calculate to two significant figures the distances between the points

   a   (1, 1) and (7, 4)   *b* $(-3, 4)$ and $(-2, 6)$.

45   A ship sails 10 km SE and then sails SW. When it is 14 km from its starting point, how far has it sailed on the second leg of the journey?

Exercise B

1   64 cm of wire are used to make a skeleton cuboid which is 7 cm long and 5 cm broad. Assuming that no wire is wasted, what is the height of the cuboid?

2   Plot the points O (0, 0), A (8, 0), B (6, 4) and C (2, 4) on squared paper. What is the equation of the axis of symmetry of the quadrilateral OABC?

      What other kind of quadrilateral has exactly one axis of symmetry?

3   What is the size of the smaller angle between the hands of a clock when the time shown is:

   a   half past 7   *b*   half past 10   *c*   20 minutes past 8?

4 *a*   Draw a triangle ABC with AB = 5·5 cm, BC = 6·5 cm and CA = 7·8 cm. Measure its angles. How can you check the accuracy of your measurements?

   b   In triangle PQR, PQ = 5·5 cm and QR = 6·5 cm. Between which lengths must RP lie?

5   The angles of elevation of the top of a tower from two points A and B on level ground on the same side of the tower and in a straight line with it are 41° and 33° respectively. If A is 20 metres further from the tower than B, make a scale drawing and find the height of the tower.

      Suggest why it is difficult to get a good answer by drawing in this case.

Cumulative Revision Exercises

6   A tree is 7 m high. From two points A and B, which are in line with the foot of the tree but on opposite sides of it, the angles of elevation are 35° and 40° respectively. Make a scale drawing, and hence find the distance between A and B.

7 *a* With rectangular axes plot the points A (2, 1), B (5, 1) and C (5, 5). Write down the coordinates of D, the fourth vertex of the rectangle ABCD.
   b What are the coordinates of M, the point of intersection of AC and BD?
   c Which of the following equations is satisfied by the coordinates of both A and C?
      (*1*) $4y = 3x-2$ (*2*) $3y = 4x-5$ (*3*) $4y = 3x+5$

8   A cube has the same volume as a cuboid with dimensions 16 cm by 2 cm by 2 cm. What is the length of an edge of the cube?

9   A is the point (6, 6) and B is (3, 2). AB and AC are the equal sides of an isosceles triangle ABC and BC is parallel to the *y*-axis. Find the coordinates of C.

10   On squared paper draw an isosceles right-angled triangle with an area of 4 square units.

11*a* On squared paper plot the points P (2, 5) and Q (2, 11).
  b Write down the coordinates of S, the midpoint of PQ.
  c Write down the coordinates of two points R and T such that triangles PRQ and PTQ are isosceles with RS = ST = 2 units.
  d Calculate the area of triangle PRQ.
  e What kind of figure is PRQT?

12   Under reflection in the line with equation $x = 1$, triangle ABC is mapped to triangle $A_1B_1C_1$.
  a If A is (2, 3), B is (3, 2) and C is (6, 6), find the coordinates of the images A_1, B_1 and C_1.
  b Reflect the points A_1, B_1 and C_1 in the origin, and state the coordinates of their images A_2, B_2 and C_2.
  c State the translation under which A would be mapped to A_2. Would this translation map B to B_2 and C to C_2?

13*a* On squared paper plot the points A (0, 6), B (2, 0) and C (4, 6).
  b Write down the coordinates of the point D such that ABCD is a rhombus.
  c Calculate the area of ABCD.

d If (x, y) is a member of the set of points on the line AC, what can you say about y?

14a Mark the points D (3, 1) and E (3, 6) in the coordinate plane.
  b Draw a kite ABCD in which E is the point of intersection of the diagonals AC (parallel to the *x*-axis) and BD, where AC = 4 units and BD = 8 units.
  c Write down the coordinates of A, B and C.
  d Calculate the area of ABCD.
  e If A and C remain fixed and the kite is turned over, draw the new position of the kite and state the coordinates of B_1 and D_1, the new positions of B and D.

15a Plot the points A (3, 10), B (2, 3) and C (−5, 4) in the coordinate plane.
  b What are the coordinates of D if ABCD is a parallelogram?
  c Find the coordinates of the point about which the parallelogram is conserved by a half turn.

16a A is the point (3, 1), B is (7, 1), D is (5, −3) and E is (1, −3). Illustrate by a diagram.
  b What kind of figure is ABDE? Calculate its area.
  c DBPQ is congruent to ABDE and lies on the opposite side of DB. Write down the coordinates of P and Q. What kind of figure is APQE?
  d If ABDE is reflected in AB, write down the coordinates of the images of D and E.

17 In Figure 8 AB, DE and GH are parallel, and AG and BH are parallel.
  a Name 4 pairs of corresponding angles.
  b Name 4 pairs of alternate angles.
  c How many different sizes of corresponding angles are there?
  d How many different sizes of alternate angles are there?

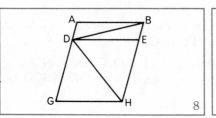

8

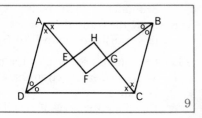

9

Cumulative Revision Exercises

18 In Figure 9 ABCD is a parallelogram and the bisectors of its angles cut in E, F, G and H. Explain why angle AED = 90°. What kind of shape is EFGH?

19a Show by different types of shading on a diagram the sets
$A = \{(x, y): y \leq 5\}$, $B = \{(x, y): y \geq x\}$ and $C = \{(x, y): x \geq 0\}$.
b Describe $A \cap B \cap C$ by completing $A \cap B \cap C = \{(x, y): \ldots\}$.
c What is the shape of the region given by $A \cap B \cap C$? Calculate its area.

20 Illustrate, in separate diagrams, the sets:
a $P \cap Q$ b $Q \cap R$
where $P = \{(x, y): x \leq 2\}$, $Q = \{(x, y): y < 2x\}$ and $R = \{(x, y): y < 3 - 2x\}$.

21 A quadrilateral ABCD is given a translation $\begin{pmatrix} -5 \\ 5 \end{pmatrix}$ which maps it to $A_1B_1C_1D_1$. State as much as you can about $\vec{AA_1}$, $\vec{BB_1}$, $\vec{CC_1}$ and $\vec{DD_1}$ in respect of:
a direction b magnitude.

22 Calculate the values of p and q in each of the given translation equations:

a $\begin{pmatrix} 5 \\ -3 \end{pmatrix} + \begin{pmatrix} 2 \\ 5 \end{pmatrix} + \begin{pmatrix} -4 \\ -6 \end{pmatrix} = \begin{pmatrix} p \\ q \end{pmatrix}$ b $\begin{pmatrix} 12 \\ -1 \end{pmatrix} + \begin{pmatrix} p \\ q \end{pmatrix} + \begin{pmatrix} -10 \\ 4 \end{pmatrix} = \begin{pmatrix} 0 \\ 0 \end{pmatrix}$

23 If $\vec{AB} \oplus \vec{BC} \oplus \vec{CA} = 0$, which of the following statements is true?
a ABC must be a triangle.
b A, B and C must be in the same straight line.
c Either ABC is a triangle, or A, B and C are in the same straight line.

24 In triangle ABC, AD is perpendicular to BC. If AB = 26 cm, AC = 12·5 cm and AD = 10 cm, calculate the length of BC. (Consider two cases: $\angle$ACB acute, and $\angle$ACB obtuse.)

25 An aeroplane flies 24 km due east from A to B and then flies due north to C. If it can return from C to A in 6 minutes, flying at 400 km/h, how far is C from B?

Geometry

26 Under a translation $\binom{5}{6}$ the point A (4, 8) is mapped to A_1. A translation $\binom{3}{0}$ then maps A_1 to A_2.

a Give the coordinates of A_1 and A_2.
b Calculate the lengths of AA_1, A_1A_2 and AA_2.
c If A is mapped to A_2 under a single translation $\binom{a}{b}$, what are the values of a and b?

27 In Figure 10, use the converse of Pythagoras' theorem to show that the given measurements are possible in triangle ABC. (Remember that two right angles make up a straight angle.)

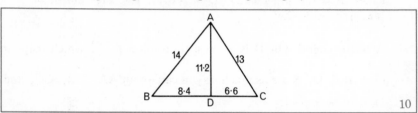

28 ABCD is a rectangle with AB = 12 km and BC = 7 km. M is the midpoint of CD and N is a point on CD 4 km from C.
Show that AM + MB < AN + NB < AC + CB.
Repeat the exercise for any other position of N between M and C.

29a Draw a rhombus of side 6 cm if the rhombus is formed from two equilateral triangles.
b Draw the diagonals of the rhombus and mark the size of each angle in the resulting figure.
c Calculate the length of the longer diagonal and also the area of the rhombus.

30 In an isosceles triangle ABC, AB = AC = 61 cm and BC = 22 cm. If AD is the altitude from A, calculate:

a the length of AD *b* the area of $\triangle ABC$
c the length of the perpendicular from B to AC.

31 A translation of magnitude 13 units is represented by the number pair $\binom{p}{-5}$. What are the possible replacements for p?
What property of the translation do we not know in this case?

Cumulative Revision Exercises 157

32 The join of the midpoints M and N of AB and DC is an axis of symmetry of the quadrilateral ABCD. In the figure, name:
  a three pairs of equal lines   *b* right angles
  c two pairs of equal angles which are not necessarily right
  d two parallel lines   *e* four pairs of supplementary angles.

33 A kite PQRS has vertices P $(-2, 2)$, Q $(1, -5)$, R $(4, 2)$ and S $(1, k)$ where k is a real number.
  Which replacements for k:
  a are not possible   *b* will make PQRS a rhombus
  c will make the area equal to 30 square units
  d will make the second diagonal half the length of PR?

Arithmetic

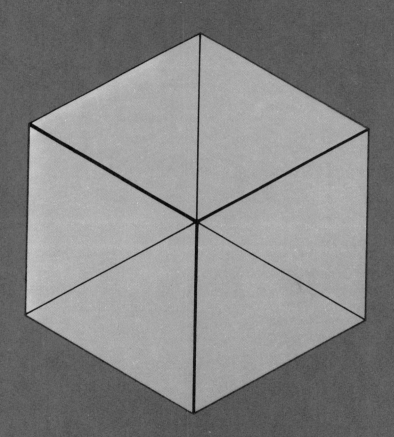

Arithmetic

Note to the Teacher on Chapter 1

Chapter 1 is divided into two main sections. In *Section* 1 the *squares* of numbers are found by direct calculation, and then with the aid of a graph, a table of squares, and a slide rule. In *Section* 2 there is a roughly parallel treatment of the inverse operation—finding *square roots*. Not all methods need be given to all pupils.

In the examples, point is given to those two processes by frequent reference to the length of the side and the area of a square.

Section 1. **Squares**

(i) Pupils should be able to recognize the squares of the numbers 0, 1, 2, 3, ..., 20 and hence to find quickly $10^2, 20^2, \ldots$ and $(0\cdot1)^2, (0\cdot2)^2, \ldots$ A thorough understanding of this will help later to avoid errors of the type $\sqrt{0\cdot9} = 0\cdot3$.

Better pupils may be introduced to the quick method of finding (say) $3\cdot5^2$ and 75^2 by writing

$$(3\cdot5)^2 = (3\times4) + 0\cdot25 = 12\cdot25$$
$$[\text{compare } (n+\tfrac{1}{2})^2 = n^2 + n + \tfrac{1}{4} = n(n+1) + \tfrac{1}{4}]$$
and $\quad (75)^2 = (7\times8)\times100 + 25 = 5625.$

(ii) Reading off squares from a *graph* is a useful exercise in itself, and links up with the idea of a mapping in Algebra. On the graph in Figure 1, note the different scales on the two axes and the effect this has on the accuracy to be expected. On the horizontal axis the 'numbers' can be estimated to the nearest 0·01, but their 'squares' can be expected only to the nearest 0·1.

(iii) Throughout this series of books, *three-figure tables** are used; these will give results as good as those obtained by slide rule, and avoid difficulties and errors resulting from difference columns. Exercise 3 is restricted to numbers between 1 and 10, and this may well be sufficient for some pupils. Exercise 4B extends the process to numbers greater than 10 or less than 1 by first writing them in standard form.

(iv) Squares of numbers can be found quickly using a slide rule, as explained in this part of the chapter.

---

* *Three-figure Tables for Modern Mathematics* (Blackie/Chambers)

Section 2. **Square roots**

(i) Finding the square root of a number is introduced as the *inverse operation* of squaring a number. Here we consider only non-negative numbers and square roots.

Note. $N \to N^2$ and $N^2 \to N$ are inverse operations when we are dealing with non-negative numbers. Hence we take $\sqrt{N}$ ($N \geqslant 0$) to be that non-negative number whose square is N. This is, of course, the non-negative root of the equation $x^2 = N$, which has solution set $\{\sqrt{N}, -\sqrt{N}\}$. For example, $\sqrt{49} = 7$, and the roots of the equation $x^2 = 49$ are 7 and -7. The important point in this chapter is that $\sqrt{49}$ means $7(-\sqrt{49}$ would be -7.)

In Exercise 7, the numbers are restricted to exact square roots; it is suggested that many more easy oral examples be given to consolidate the basic meaning before proceeding further. These examples might well include some in which the side of a square has to be found when the area is given.

(ii) In introducing the use of the *graph* in Figure 1 attention should be drawn to the direction of the arrows in the 'guiding' lines.

Horizontal axis $\to$ vertical axis gives the *square*, e.g. $5^2 = 25$
Vertical axis $\to$ horizontal axis gives the *square root*,
$$\text{e.g. } \sqrt{49} = 7$$

Here, for the first time, pupils are faced with the problem of estimating the square root of a number which is not an exact square. The fact that (say) $\sqrt{3}$ is irrational may be discussed informally thus: 'What is the square root of 3?'

Probably the first answer suggested will be $1\frac{1}{2}$.
'Let us check this:

$$(1\tfrac{1}{2})^2 = \tfrac{3}{2} \times \tfrac{3}{2} = \tfrac{9}{4} \neq 3 \quad \text{so} \quad \sqrt{3} \neq 1\tfrac{1}{2}.$$

In the same way,

$$(1\tfrac{3}{4})^2 = \tfrac{7}{4} \times \tfrac{7}{4} = \tfrac{49}{16} \neq 3 \quad \text{so} \quad \sqrt{3} \neq 1\tfrac{3}{4}.$$

We see that $\tfrac{3}{2}$ cannot be simplified by dividing the numerator and denominator by a common factor, neither can $\tfrac{3}{2} \times \tfrac{3}{2}$. So there is no way in which $\tfrac{3}{2} \times \tfrac{3}{2}$ can give a whole number. The same reasoning applies to any other answer (e.g. $\tfrac{7}{4}$) which is a fraction. But since 3 lies between 1 and 4, $\sqrt{3}$ lies between 1 and 2, and $\sqrt{3}$ cannot be expressed as an exact fraction.

$\sqrt{3}$ is an *irrational* number.

(iii) Finding a *good estimate* of the square root of a number is an essential part of this course. The method suggested here is equivalent to a linear interpolation. It will be found that most pupils can soon make a good estimate without drawing the two lines suggested. For example, to estimate $\sqrt{84}$ we have $\sqrt{81} = 9$ and $\sqrt{100} = 10$; half-way between 81 and 100 is about 90, and 84 is nearer 81 than 90; so a good estimate might be 9·2. However, 9·1 and 9·3 would be acceptable estimates.

(iv) Exercise 10 gives practice in using *tables* to find the square roots of numbers, first of all in the range 1 to 10, then in the range 10 to 100, and lastly in a combination of these two ranges. Since most of the numbers commonly encountered fall within the range 1 to 100, it is suggested that less able pupils might well omit the section on numbers greater than 100 or less than 1. Even the better pupils will find some difficulty here, and should always check their square roots by rough estimates.

(v) Making a rough estimate of the answer is essential when using the *slide rule*, and this should be insisted on in finding square roots.

(vi) The *iterative method* of finding square roots has been included for the following reasons:

  (1) This method can be justified to the pupil more easily than the traditional method.

  (2) In practice, the modern method of calculating square roots to a high degree of accuracy is either to find a first estimate from slide rule or tables and then to use a calculating machine to iterate, or to use an iterative method with a computer.

  (3) The process of iteration can be applied to other problems, such as finding the solutions of certain equations.

By considering pairs of numbers ('factors') whose product is 36, we see that in each pair one factor must be less than, and the other greater than, the square root, 6. We also see that, in the example given, the average of 18 and 2 is 10, of 9 and 4 is 6·5, of 8 and 4·5 is 6·25. We notice that, as the first factors (18, 9, 8) get closer and closer to the square root, the averages (10, 6·5, 6·25) get closer still to the square root. This suggests the method of Example 1 on page 171.

Pupils have learned how to make a good estimate of the square root, so it seems sensible to choose this as their first 'factor'. They get the second 'factor' by division, and find the average of these 'factors'. This ought to give a still closer approximation to the square root. A few examples done with a good estimate followed by one division will show pupils that results so obtained are very close to the values given in the three-figure tables.

We have already seen that the closer the first 'factor' is to the square root the closer will be the average of the two 'factors'. Hence if we want an even closer approximation we can take as our second estimate the result of the first iteration and repeat the process. It may be useful to have a book of square roots to 7 or 10 figures available in order to verify that the statement in the last sentence is true. Examples 2 and 3 suggest a compact way of setting down the working.

Justification of the method (for the Teacher)

Let N be a number between 1 and 100.

Let $a_1 =$ first estimate to $\sqrt{N}$ and $q_1 = \dfrac{N}{a_1}$.

$a_2 = \frac{1}{2}(a_1 + q_1) =$ second estimate, $q_2 = \dfrac{N}{a_2}$, and so on.

Since for any two positive numbers the arithmetic mean is greater than the geometric mean, $a_2 > \sqrt{N}$, so every estimate after the first is greater than $\sqrt{N}$.

If we assume $a_2 > \sqrt{N}$ it follows that

$$a_1 > a_2 > \sqrt{N} > q_2 > q_1$$

and generally $a_2 > a_3 > a_4 > \ldots > \sqrt{N}$ and each estimate is nearer the square root than the previous one.

If the error in estimating $\sqrt{N}$ as a_1 is h, we have

$$\sqrt{N} = a_1 - h$$
$$\Rightarrow \quad N = a_1^2 - 2a_1 h + h^2$$
$$\Rightarrow \quad \frac{N}{a_1} = a_1 - 2h + \frac{h^2}{a_1} = q_1$$

$$\Rightarrow \tfrac{1}{2}(a_1+q_1) = \tfrac{1}{2}a_1 + \tfrac{1}{2}\left(a_1 - 2h + \frac{h^2}{a_1}\right)$$
$$= a_1 - h + \frac{h^2}{2a_1} = \sqrt{N} + \frac{h^2}{2a_1}$$

i.e. $\quad a_2 = \sqrt{N} + \dfrac{h^2}{2a_1}$

$\Rightarrow \quad \sqrt{N} = a_2 - \dfrac{h^2}{2a_1}$

If h is small, say 0.1, the error in a_2, $\dfrac{h^2}{2a_1}$, will be $\dfrac{0.01}{2a_1}$ which is certainly less than 0.005. So if we start with a good estimate, one division should give a result correct to 3 significant figures.

If h is not small we may have a considerable number of divisions to do before reaching a reasonable degree of accuracy. For example, if in calculating $\sqrt{99}$ we start with $a_1 = 1$, five divisions are necessary before we reach the estimate $a_5 = 9.97$, which is not even correct to 3 significant figures. This does not matter if we are using an electronic computer so it is common practice for a computer programmer to take 1 as the first estimate.

In general, if one iteration is correct to n significant figures ($n \geqslant 2$), the next will be correct to $2n-1$ significant figures.

(facing page 161)

Squares and Square Roots of Numbers

1 Squares of numbers

(i) Calculating squares of numbers

In Book 1 you used the set of squares of whole numbers:

$$\{0^2, 1^2, 2^2, 3^2, \ldots\}$$
$$= \{0, 1, 4, 9, \ldots\}$$

We can find the square of any number by multiplying the number by itself.

Examples

1. 15^2
   $= 15 \times 15$
   $= 225$

2. $0 \cdot 2^2$
   $= 0 \cdot 2 \times 0 \cdot 2$
   $= 0 \cdot 04$

3. $2 \cdot 5^2$
   $= 2 \cdot 5 \times 2 \cdot 5$
   $= 6 \cdot 25$

4. Area of square of side 6 m
   $= 6^2$ m^2
   $= 36$ m^2

Exercise 1

1. Write down the squares of all the whole numbers from 0 to 10.
2. Calculate the squares of all the whole numbers from 11 to 20.
3. Calculate 25^2, 30^2, 50^2 and 75^2.
4. Calculate the squares of 1·5, 4·5, 7·5 and 9·5.
5. Find the squares of 0·1, 0·3, 0·5, 0·7 and 0·9.
6. Write down the squares of 2, 20, and 200.
7. Write down the values of 100^2, 300^2, 400^2 and 500^2.
8. Find the areas of the squares with sides:
   4 cm, 7 m, 2 km, 2·1 m, 0·8 km and 4·7 cm.

Arithmetic

(ii) Using a graph

In Figure 1 the dots show the graph of the mapping $x \to x^2$ for $x = 0, 1, 2, \ldots, 10$

x	0	1	2	3	4	5	6	7	8	9	10
x^2	0	1	4	9	16	25	36	49	64	81	100

By joining the points on this graph by a smooth curve we can relate all numbers from 0 to 10 with their squares.

Notice the arrows which indicate $5^2 = 25$ and $6 \cdot 2^2 \doteqdot 38 \cdot 4$.

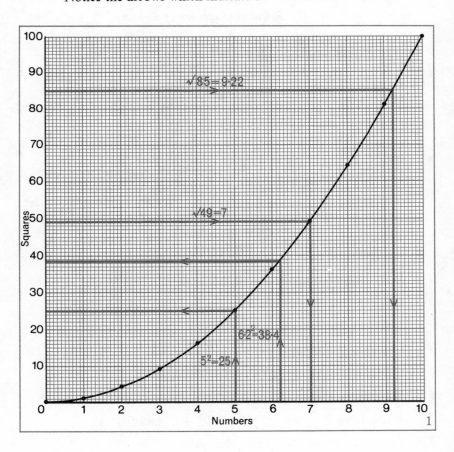

Squares of numbers

Exercise 2

1. Use the graph in Figure 1 to write down the squares of the following numbers (to 1 decimal place where appropriate):

   a 8   b 9   c 4·5   d 2·3   e 7·1   f 6·5
   g 7   h 10 i 3·3   j 5·6   k 8·2   l 9·5

2. Use the graph to find the squares of several numbers between 0 and 10 of your own choice, then check your answers by *calculating* the squares.

(iii) Using tables of squares

Tables of squares of numbers from 1 to 10 are given on pages 264 and 265.

To find $5·57^2$ we use the part of the table shown below.

	0	1	2	3	4	5	6	7	8	9
5·5	30·25	30·36	30·47	30·58	30·69	30·80	30·91	31·02	31·14	31·25

We find the first two figures of 5·57 in the left-hand column of the table, and the third figure at the top of the page. So we read across from 5·5 until we come to the number below 7. Thus $5·57^2 = 31·02$.

Similarly $5·5^2 = 30·25$, and $5·59^2 = 31·25$.

Notice that these tables normally give the squares rounded off to 2 decimal places.

Exercise 3

Use the tables of squares on pages 264 and 265 to write down the squares of the following numbers to 2 decimal places:

1 2·5   2 3·5   3 4·5   4 4·51   5 4·52
6 4·59   7 7·0   8 7·01   9 7·10   10 9·2
11 9·29 12 6·99 13 2·02 14 5·47 15 8·76

*   *   *   *

Arithmetic

To find the squares of numbers greater than 10 or less than 1

Example 1. Use tables to find $45 \cdot 6^2$.
The tables give the squares of numbers from 1 to 10 only.
So we write $\quad 45 \cdot 6 = 4 \cdot 56 \times 10 \quad\quad (4 \cdot 56$ lies between 1 and 10)
Hence $\quad\quad 45 \cdot 6^2 = (4 \cdot 56 \times 10)^2$
$\quad\quad\quad\quad\quad = 4 \cdot 56^2 \times 10^2$
$\quad\quad\quad\quad\quad = 20 \cdot 79 \times 100$
$\quad\quad\quad\quad\quad = 2079$

Example 2. $\quad 139^2 = (1 \cdot 39 \times 100)^2$
$\quad\quad\quad\quad\quad = 1 \cdot 39^2 \times 100^2$
$\quad\quad\quad\quad\quad = 1 \cdot 93 \times 10000$
$\quad\quad\quad\quad\quad = 19300$

Example 3. $0 \cdot 78^2 = \left(\dfrac{7 \cdot 8}{10}\right)^2$
$\quad\quad\quad\quad = \dfrac{60 \cdot 84}{100}$
$\quad\quad\quad\quad = 0 \cdot 6084$

Exercise 4B

Use tables and the above method to find the squares of:

1 *a* $23 \cdot 5$ *b* $37 \cdot 1$ *c* $89 \cdot 2$ *d* $10 \cdot 1$

2 *a* 456   *b* 209   *c* 500   *d* 905

3 *a* $0 \cdot 29$ *b* $0 \cdot 87$ *c* $0 \cdot 56$ *d* $0 \cdot 33$

Find the values of the following:

4 *a* $12 \cdot 9^2$ *b* 152^2 *c* $0 \cdot 78^2$ *d* $0 \cdot 789^2$

5 *a* $30 \cdot 7^2$ *b* $123 \cdot 8^2$ *c* $653 \cdot 2^2$ *d* $0 \cdot 206^2$

(iv) Using a slide rule

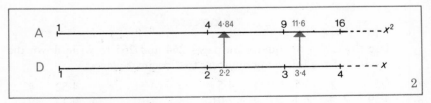

On a slide rule the numbers on the A scale (x^2) are the squares of the numbers on the D scale (x) which are in alignment with them.

Squares of numbers

The method, therefore, is to place the line on the cursor over the number on the D scale and to read off its square under the line on the A scale. Figure 2 illustrates that

$$2 \cdot 2^2 = 4 \cdot 84, \quad \text{and that} \quad 3 \cdot 4^2 \doteqdot 11 \cdot 6.$$

Note: 1. A slide rule will normally give an approximation to 3 significant figures.
2. Since both the D and A scales are on the stock of the rule, removing the 'slide' may help to avoid errors.
3. On some slide rules the D scale is above the A scale.

Exercise 5

Using your slide rule, write down the squares of the following numbers as accurately as you can:

	a		b		c		d		e	
1		1·5		2·4		3·1		1·55		2·45
2		4·5		5·6		7·8		9·45		6·22
3		2·12		3·75		1·83		8·25		9·1

Exercise 5B

	a		b		c		d		e	
1		1·1		11		6·3		63		630
2		2·7		27		270		0·27		2·07
3		1·73		0·84		73·8		3·05		123

Exercise 6 (Miscellaneous)

1   Use any of the above methods to find the squares of:

a	5	b	15	c	25	d	50	e	500
f	8·5	g	0·5	h	12·5	i	17	j	71
k	0·39	l	0·246	m	128	n	271·9	o	15·8

2   Calculate the areas of squares with sides of lengths:

    a 3·5 cm   b 16 mm   c 1·06 m   d 37·1 cm

3   Find the values of:

    a $3^2 + 13^2$ b $2 \cdot 3^2 + 1 \cdot 7^2$ c $8 \cdot 4^2 - 1 \cdot 6^2$

Arithmetic

2 Square roots of numbers

(i) Calculating square roots of numbers

Just as *addition* and *subtraction* of numbers are inverse operations, and *multiplication* and *division* of numbers are inverse operations, so *squaring* and *taking the square root* of numbers are *inverse operations*.

Number	Square	Number	Square root
0 →	0	0 →	0
1 →	1	1 →	1
2 →	4	4 →	2
3 →	9	9 →	3
4 →	16	16 →	4
.	.	.	.
.	.	.	.
.	.	.	.

Thus since the square of 3 is 9, the square root of 9 is 3.
Just as we write $3^2 = 9$, so we write $\sqrt{9} = 3$.
Similarly $\sqrt{4} = 2$, $\sqrt{1} = 1$, $\sqrt{0} = 0$, and $\sqrt{6\cdot25} = 2\cdot5$.

The positive square root of a positive number N is that number $\sqrt{N}$ which, when multiplied by itself, gives N. The square root of 0 is 0.

Exercise 7

1 Write down the square roots of:
  a 1   *b* 16   *c* 36   *d* 64   *e* 100   *f* 400

2 Write down the values of:
  a $\sqrt{25}$ *b* $\sqrt{49}$ *c* $\sqrt{81}$ *d* $\sqrt{144}$ *e* $\sqrt{900}$ *f* $\sqrt{1600}$

3 Given $289 = 17^2$, copy and complete $\sqrt{289} = \ldots$

4 Given $529 = 23^2$, copy and complete $\sqrt{529} = \ldots$

5 Given $12\cdot25 = 3\cdot5^2$, copy and complete $\sqrt{12\cdot25} = \ldots$

6 Given $1\,000\,000 = 1000^2$, copy and complete $\sqrt{1\,000\,000} = \ldots$

7 How long is the side of a square whose area is:
  a 9 cm²   *b* 36 m²   *c* 100 km²   *d* 225 cm²   *e* 1·44 m²?

Square roots of numbers

(ii) Using a graph

Notice the arrows on the graph in Figure 1 (page 162) which indicate the operation of finding a square root:

$$\sqrt{49} = 7 \quad \text{and} \quad \sqrt{85} \doteqdot 9\cdot 22$$

Exercise 8

1. Use the graph in Figure 1 to write down the values of the following to 1 decimal place:

   a $\sqrt{20}$ b $\sqrt{30}$ c $\sqrt{40}$ d $\sqrt{50}$ e $\sqrt{60}$ f $\sqrt{70}$

2. Use the graph to find, to 1 decimal place, the square roots of:

   a 15 b 55 c 75 d 88 e 90 f 37

3. Use the graph to find approximately the lengths of the sides of squares whose areas are:

   a 19 cm² b 54 m² c 74 mm² d 96 m² e 28 cm² f 80 mm²

(iii) Estimating square roots of numbers

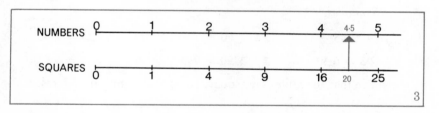

From the number line and the line of squares in Figure 3 we see that $\sqrt{20} \doteqdot 4\cdot 5$.

We can check this estimate by squaring 4·5.
$4\cdot 5^2 = 20\cdot 25$, which shows that 4·5 is a good approximation for $\sqrt{20}$.

Exercise 9

1. Between which two consecutive whole numbers do the square roots of the following numbers lie?

   a 5 b 24 c 87 d 137 e 13·7 f 47

Arithmetic

2. *Estimate* the square root of each of the following numbers to 1 decimal place, and then check your answer by squaring:
   a 42   b 70   c 94   d 8   e 5   f 30

3. *Estimate* the following to 2 significant figures:
   a $\sqrt{12}$ b $\sqrt{50}$ c $\sqrt{3}$ d $\sqrt{7}$ e $\sqrt{90}$ f $\sqrt{6}$

4. *Estimate* $\sqrt{27}$ and $\sqrt{2\cdot7}$, and check by squaring. Note the importance of the position of the decimal point.

5. *Estimate*   a $\sqrt{6}$ and $\sqrt{60}$ b $\sqrt{47}$ and $\sqrt{4\cdot7}$
                c $\sqrt{96}$ and $\sqrt{9\cdot6}$ d $\sqrt{1\cdot4}$ and $\sqrt{14}$

(iv) Using tables of square roots

Tables of square roots of numbers from 1 to 10 are given on pages 266 and 267.

Tables of square roots of numbers from 10 to 100 are given on pages 268 and 269.

Caution: Make sure you look up the correct table.

Examples

1. $\sqrt{6} = 2\cdot45$ 2. $\sqrt{60} = 7\cdot75$ 3. $\sqrt{4\cdot23} = 2\cdot06$
4. $\sqrt{42\cdot3} = 6\cdot50$ 5. $\sqrt{90} = 9\cdot49$ 6. $\sqrt{92\cdot8} = 9\cdot63$

Notice that the tables give square roots to 3 significant figures.

Exercise 10

1. Use the tables of square roots to find the square roots of the following, to 3 significant figures:
   a 5   b 5·01   c 5·07   d 5·10   e 5·11   f 5·18
   g 6·8   h 7·65   i 9·04   j 1·01   k 3·5   l 6·82

2. From the tables find the square roots of:
   a 10   b 30   c 30·6   d 36·0   e 52·9   f 87·6

3. Use tables to find:
   a $\sqrt{77}$ b $\sqrt{22\cdot2}$ c $\sqrt{2}$ d $\sqrt{3\cdot81}$ e $\sqrt{54\cdot9}$ f $\sqrt{6\cdot41}$

Square roots of numbers

4 Calculate the lengths of the sides of squares with areas:
   a   50 cm^2 *b*   16·3 mm^2 *c*   88 cm^2 *d*   1·75 m^2 *e*   7·05 mm^2

5 Find the values of:
   a $\sqrt{(1^2+2^2+3^2)}$ *b* $\sqrt{(13^2-5^2)}$ *c* $\sqrt{(2\cdot1^2-1\cdot2^2)}$

*   *   *   *

To find the square roots of numbers greater than 100 or less than 1

Examples

1. Use tables to find $\sqrt{123}$.
The tables give square roots of numbers from 1 to 100 only, so $\sqrt{123}$ cannot be found directly.
The tables do show us, however, that
$\sqrt{1\cdot23} = 1\cdot11$ and $\sqrt{12\cdot3} = 3\cdot51$
Hence
EITHER $\sqrt{123} = \sqrt{(1\cdot23 \times 100)}$ OR $\sqrt{123} = \sqrt{(12\cdot3 \times 10)}$
              $= \sqrt{1\cdot23} \times \sqrt{100}$ $= \sqrt{12\cdot3} \times \sqrt{10}$
              $= 1\cdot11 \times 10$ $= 3\cdot51 \times 3\cdot16$
              $= 11\cdot1$ $\doteqdot 11\cdot09 \doteqdot 11\cdot1$

Since 100 has an exact square root it is clear that the *first* of these alternatives is much simpler.

To find the square root of a number between 100 and 10 000 first express the number as a multiple of 100.

2. $\sqrt{6020} = \sqrt{(60\cdot2 \times 100)} = 7\cdot76 \times 10 = 77\cdot6$
3. $\sqrt{193\cdot6} \doteqdot \sqrt{(1\cdot94 \times 100)} = 1\cdot39 \times 10 = 13\cdot9$

A similar method, illustrated in the following two examples, is used for numbers less than 1.

4. $\sqrt{0\cdot123} = \sqrt{\dfrac{12\cdot3}{100}} = \dfrac{\sqrt{12\cdot3}}{10} = \dfrac{3\cdot51}{10} = 0\cdot351$

5. $\sqrt{0\cdot0123} = \sqrt{\dfrac{1\cdot23}{100}} = \dfrac{\sqrt{1\cdot23}}{10} = \dfrac{1\cdot11}{10} = 0\cdot111$

Arithmetic

To find the square root of a number between 1 and 0·01 first express the number as a multiple of $\frac{1}{100}$.

Exercise 11B

Write the numbers in questions *1* to *3* in the form $a \times 100$ or $a \times \frac{1}{100}$ so that a is a number between 1 and 100.

	a		b		c		d		e	
1	a	234	b	638	c	200	d	2135	e	3047
2	a	0·52	b	0·63	c	0·5	d	0·145	e	0·204
3	a	0·06	b	0·025	c	0·7	d	0·0246	e	0·7777

Use tables to find the square roots of the numbers in questions *4* to *7*.

	a		b		c		d		e	
4	a	135	b	872	c	407	d	500	e	567·8
5	a	1230	b	5900	c	8800	d	2488	e	278
6	a	0·36	b	0·88	c	0·789	d	0·45	e	0·045
7	a	0·0731	b	0·018	c	432	d	0·917	e	3726

8 Find the values of:

	a	$\sqrt{6500}$	b	$\sqrt{2\cdot34}$	c	$\sqrt{804}$	d	$\sqrt{36\cdot92}$	e	$\sqrt{0\cdot753}$
	f	$\sqrt{0\cdot06}$	g	$\sqrt{2300}$	h	$\sqrt{12\cdot5}$	i	$\sqrt{0\cdot9753}$	j	$\sqrt{498}$

The above methods can be extended to numbers greater than 10000 or less than 0·01.

e.g. $\sqrt{19360} = \sqrt{(1\cdot94 \times 10000)} = 1\cdot39 \times 100 = 139$

$$\sqrt{0\cdot0039} = \sqrt{\frac{39}{10000}} = \frac{6\cdot24}{100} = 0\cdot0624$$

Find the square roots of:

	a		b		c		d		e	
9	a	13500	b	24600	c	123456	d	134000	e	67543
10	a	0·008	b	0·0076	c	0·0005	d	0·0008	e	0·0029

(v) Using a slide rule

You have already used your slide rule to read from the A scale the square of any number on the D scale (page 164).

The inverse operation, A scale → D scale, gives the square roots of numbers.

Square roots of numbers

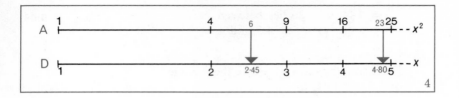

Figure 4 illustrates that $\sqrt{6} = 2\cdot 45$ and that $\sqrt{23} = 4\cdot 80$.

Note: An estimate should be made when using a slide rule to find square roots.

Exercise 12

Use your slide rule to find the square roots of the following to 3 significant figures.

1	*a*	3	*b*	8	*c*	30	*d*	68	*e*	95	*f*	48
2	*a*	4·4	*b*	7·6	*c*	9·8	*d*	35·4	*e*	82·8	*f*	5·32
3	*a*	183	*b*	675	*c*	2890	*d*	7480	*e*	852·9	*f*	45·2
4	*a*	0·02	*b*	0·20	*c*	0·78	*d*	0·078	*e*	0·384	*f*	0·0384
5	*a*	192	*b*	0·678	*c*	8767	*d*	0·097	*e*	0·456	*f*	432·8

(vi) Using an iterative method

If we divide 36 by 6 the quotient is 6, so $\sqrt{36} = 6$.
Also, $\frac{36}{2} = 18$, so $\sqrt{36}$ lies between 2 and 18
$\frac{36}{3} = 12$, so $\sqrt{36}$ lies between 3 and 12
$\frac{36}{4} = 9$, so $\sqrt{36}$ lies between 4 and 9

Thus if 36 is divided by a number *a* to give a quotient *b*, $\sqrt{36}$ lies between *a* and *b*.

This is the basis for the following method of finding square roots.

Example 1. Calculate $\sqrt{20}$
A first estimate = 4·5 (from number line).
If we divide 20 by 4·5 and get 4·5 exactly as the quotient, then this is the square root of 20. If we do not obtain 4·5 exactly, then $\sqrt{20}$ lies between 4·5 and the quotient.

Arithmetic

$\dfrac{20}{4\cdot 5} = 4\cdot 44$, so $\sqrt{20}$ lies between 4·5 and 4·44

$\dfrac{20}{4\cdot 5} = \dfrac{200}{45} = \dfrac{40}{9} = 4\cdot 44$

For a second estimate we take the average of 4·5 and 4·44.

A second estimate $= \dfrac{4\cdot 5 + 4\cdot 44}{2} = 4\cdot 47$

$\dfrac{20}{4\cdot 47} = 4\cdot 474$, so $\sqrt{20}$ lies between 4·47 and 4·474

A third estimate $= \dfrac{4\cdot 47 + 4\cdot 474}{2} = 4\cdot 472$

(4·472 is in fact accurate to 4 significant figures.)

```
        4·474
447)2000·000
     1788
     2120
     1788
     3320
     3129
     1910
```

Each of the estimates 4·5, 4·47, 4·472 is a better approximation for $\sqrt{20}$ than the previous one.

This is an *iterative* method in which an approximation at one stage is used to obtain a better approximation at the next stage.

Note:
  (i) With a good first estimate an approximation to 3 significant figures can be obtained by one division (see Example 2).
  (ii) If the first approximation is taken from tables or slide rule, one division will give the answer to 5 significant figures (see Example 3).

Example 2. Calculate $\sqrt{8\cdot 25}$ to 3 significant figures.

First estimate $= 2\cdot 8$

$\dfrac{8\cdot 25}{2\cdot 8} = 2\cdot 946$

Second estimate $= \dfrac{2\cdot 8 + 2\cdot 946}{2}$

$ = 2\cdot 873$

$\sqrt{8\cdot 25} = 2\cdot 87$ to 3 significant figures.

```
       2·946
28)82·500
   56
   265
   252
   130
   112
   180
```

Square roots of numbers

Example 3. Write down the value of $\sqrt{17\cdot6}$ to 3 significant figures from tables, and then by one division find $\sqrt{17\cdot6}$ to 5 significant figures.

First estimate $= 4\cdot20$ (from tables)

$$\frac{17\cdot6}{4\cdot20} = 4\cdot19048$$

Second estimate $= \dfrac{4\cdot20 + 4\cdot19048}{2}$

$= 4\cdot19524$

$\sqrt{17\cdot6} = 4\cdot1952$, to 5 significant figures

```
       4·19048
  42)176·00000
     168
      80
      42
     380
     378
     200
     168
     320
```

Exercise 13

Use the method of iteration to find by a good first estimate and one division the square roots of the following numbers, rounded off to 3 significant figures.

1 a 6   *b* 29   *c* 88   *d* 56   *e* 40

2 a 7   *b* 90   *c* 17   *d* 20   *e* 3

Use the method of questions *1* and *2* to calculate:

3 a $\sqrt{14\cdot8}$ *b* $\sqrt{8\cdot2}$ *c* $\sqrt{75\cdot4}$ *d* $\sqrt{56\cdot7}$ *e* $\sqrt{1\cdot59}$

4 Write down the square root of each of the following to 3 significant figures using tables or slide rule, and by one division calculate the square root to 5 significant figures:

   a 12   *b* 69   *c* 8   *d* 21·6   *e* 93·4

Arithmetic

Summary

1 *Squares of numbers*

  a To find the square of a number, multiply the number by itself, e.g. $7^2 = 49$, $0 \cdot 2^2 = 0 \cdot 04$, $2 \cdot 5^2 = 6 \cdot 25$

  b Squares of numbers may be found by:
    (i) calculation
    (ii) a graph
    (iii) tables
    (iv) slide rule.

2 *Square roots of numbers*

  a Squaring, and taking the square root, are *inverse operations*.

  b The positive square root of a positive number N is that number which, when multiplied by itself, gives N, e.g. $\sqrt{49} = 7$, $\sqrt{0 \cdot 04} = 0 \cdot 2$, $\sqrt{6 \cdot 25} = 2 \cdot 5$

  c Square roots of numbers may be found by:
    (i) using the inverse operation to squaring, as above
    (ii) a graph
    (iii) estimation
    (iv) tables (If the number is greater than 100 or less than 1, it must be expressed as a multiple of 100 or $\frac{1}{100}$.)
    (v) slide rule
    (vi) an iterative method.

Note: In this chapter, we considered positive numbers only, so that, for example, $7^2 = 7 \times 7 = 49$, and $\sqrt{49} = 7$.

If, however, we consider negative numbers, we have $(-7)^2 = (-7) \times (-7) = 49$, so 49 also has -7 as a square root.

It appears then that 49 has *two* square roots, 7 and -7, i.e. $\sqrt{49} = 7$ and $-\sqrt{49} = -7$; and the square roots of 49 are $\pm\sqrt{49}$, i.e. ± 7.

This will be investigated later, in the algebra course.

Note to the Teacher on Chapter 2

Throughout the first three books of this series emphasis has been given to the laws governing the operations of addition and multiplication, and at this stage pupils should be conscious of the striking correspondence between certain of these laws as shown in *Section 2*. It appears reasonable that to the addition diagram of Figure 3(i) there should correspond a 'multiplication' diagram, and to the subtraction diagram of Figure 3(ii) there should correspond a 'division' diagram.

The scales shown in Figure 1 do not allow us to use the rule for multiplication, so we must replace these scales. Now in the *addition* scales, 0 is the identity element for addition; what is the logical replacement for 0? Surely it is the identity element for *multiplication*, that is, 1. We can verify this by looking at Figure 4; (i) shows that $a+0 = a$, and (ii) shows the exact correspondence $a \times 1 = a$.

Figure 5(i) shows a pair of strips suitable for use in addition, set for adding 1 to any number; this corresponds to Figure 4(i) when a is replaced by 1. Figure 5(ii) represents the strips which we are going to use for multiplication, and they correspond to Figure 4(ii). Now we know that the scales here have 1 as their starting point, but we are free to choose any number we wish as a replacement for p. Once we have chosen p the replacements for $q, r, \ldots$ are determined by using the principle of Figure 4(ii), namely, $p \times p = q$, $p \times q = r$, etc. If we choose to replace p by 2, then $q = 4$, $r = 8$, etc., and so the set of numbers $\{0, 1, 2, 3, \ldots\}$ in Figure 5(i) is seen to be replaced by the set $\{1, 2, 4, 8, \ldots\}$ in Figure 5(ii), and now the scales in Figure 1 become, for multiplication:

```
   1       2       4       8       16
—-|———————|———————|———————|———————|——
—-|———————|———————|———————|———————|——
   1       2       4       8       16
```

With most classes it will be worth studying in detail the practical work indicated in Exercise 2. Good pupils may study the effect of choosing, say, 3 as a replacement for p.

So far, the scales allow us to work only with numbers from the set $\{1, 2, 4, 8, \ldots\}$ and further calibration is necessary before more general multiplications can be tackled. In *Section 3* we study the mapping of the 'addition' numbers onto the 'multiplication' numbers, and by extending these sets graphically we improve the calibration. This mapping is fundamental when we come to study logarithms in Book 5.

The greatest difficulty that pupils will have in using a slide rule arises from the logarithmic form of the scales, which means that the length corresponding to a unit is different on different parts of the scale. Thus the first unit of the scale is usually broken down into tenths, and each tenth is further divided into hundredths. But farther along the scale we find the smallest unit shown to be a fiftieth, and later still the smallest unit is a twentieth. These variations can give a lot of trouble and it is strongly recommended that pupils be given plenty of practice in examples like those in question *1* of Exercise 4. A useful aid is a demonstration slide rule or an overhead projector with a transparent slide rule. If these are not available, the diagrams of Exercise 4 can be drawn on a large scale on sheets of cartridge paper which can be fixed to the classroom wall.

Section 7 on Rough Estimates is vitally important, and is generally found by the pupil to be more difficult than we would expect, particularly when the numbers lie in the range 0·5 to 1. It should be stressed that a rough estimate *must* be made every time a slide rule is used.

The problem that arises when the product lies off the scale is left till the pupils have gained a little proficiency, so up to this point we have to be careful with the choice of demonstration and drill questions; we must not demand a multiplication of the type 3×4 or a division of the type $20 \div 5$ until we have studied the technique explained in Figure 12. The difficulty can be avoided by using the A and B scales instead of the C and D scales, but this results in a loss of accuracy. Note that some slide rules have 'folded scales', which involves modification to the technique of Figure 13.

The use of a slide rule for finding squares and square roots has been referred to in Chapter 1. As always, a rough estimate is essential. Note that some slide rules have the A scale below the D scale.

The pupil should always be aware of the degree of accuracy he can hope to obtain in a slide-rule calculation. If the first digit of an answer is 1, he should have no difficulty in reading three significant figures, and might try to estimate a fourth. But if the first digit is 9, he has little hope of going much beyond two significant figures with any degree of confidence. Use of the technique of *Section* 8 helps to improve accuracy since, by starting with a division, we are able to reduce the number of settings of the rule.

Note on the mathematics of the method of approach

The mapping of *Section* 3 shows that we have a set $A = \{0, 1, 2, 3, ...\}$ with an operation $+$ on that set, and we have a set $M = \{1, 2, 4, 8, ...\}$ with an operation $\times$ on that set. If a and b are members of set A and α and β are members of set M, and if, further, $a \leftrightarrow \alpha$ and $b \leftrightarrow \beta$, then it is seen that $a+b \leftrightarrow \alpha \times \beta$.

In these circumstances we have an *isomorphism*. For more detailed information about isomorphism the teacher is referred to standard textbooks such as *Algebra and Number Systems* by Hunter, Monk, etc. (Blackie/Chambers) page 215, or *Advanced Algebra*, Part 2, by E. A. Maxwell (C.U.P.) pages 284-5.

It is interesting to realize that Napier in his invention of logarithms relied on this idea of isomorphism and did not use what has become the traditional method of deriving logarithms from indices.

(facing page 175)

Using a Slide Rule

1 Ruler addition and subtraction

Exercise 1–Practical

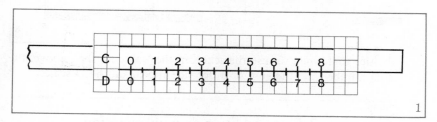

1

1. Cut two parallel slits 9 cm apart in a piece of 5-mm squared paper, as shown in Figure 1. Thread a strip of paper 1 cm wide and 20 cm long through the slits. Mark off scales on the squared paper and on the strip as shown. Call the upper scale the C scale and the lower scale the D scale. By pulling the C scale to the right we can use this device to add two numbers, as shown in Figure 2.

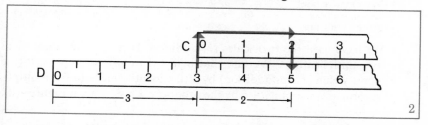

2

You can see that to add 3 and 2 we set 0 on the C scale opposite 3 on the D scale, and read off the sum on the D scale opposite 2 on the C scale.

2. Use the two scales in this way to find:

   a 4+3   *b* 5+3   *c* 2+2·5   *d* 3·5+1·5

Look at Figure 2 again and you will see that the two scales also give 5−2. The D scale simply acts as a number line.

Arithmetic

3   Use the two scales to give:

  a 8−3   *b* 5−1   *c* 4·5−2·5

Let a and b represent two numbers. The diagrams for their sum $a+b$ and for their difference $a-b$ are shown in Figure 3.

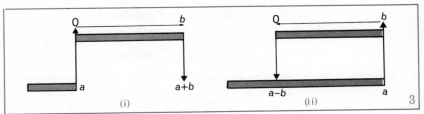

Notice that when we use the scales in the way described above, the first number and the sum or difference appear on the D scale.

2 From addition to multiplication

You have found in your study of numbers that the same laws are true for the operations of addition and multiplication. We have:

	Addition	Multiplication
	Sum $= a+b$	Product $= a \times b$
Commutative laws	$a+b = b+a$	$a \times b = b \times a$
Associative laws	$(a+b)+c = a+(b+c)$	$(a \times b) \times c = a \times (b \times c)$
Identity elements	$a+0 = a = 0+a$	$a \times 1 = a = 1 \times a$

We have used the two scales for addition. How can we adapt them for multiplication? We find that if in Figure 4(i) we replace 0 (the identity element for addition) by 1 (the identity element for multiplication), and the sum $a+b$ by the product $a \times b$ we obtain Figure 4(ii).

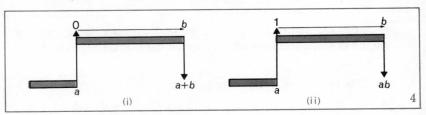

We must therefore change the number scales so that the first number is 1, as shown in Figure 5(ii). We can now choose any

Calibrating the multiplication scales

number we want as a replacement for p; suppose we take 2. It follows at once from Figure 4(ii) that 4 must be taken as the replacement for q. What number then replaces r?

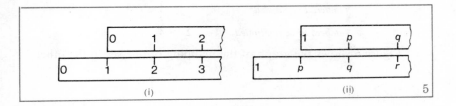

You should now be able to continue the sequence of numbers. Suppose we had replaced p by 3. What numbers would then have to be taken as replacements for q and r?

Exercise 2–Practical

1 Use the reverse sides of the squared paper and the strip you had in Exercise 1. Mark them off in centimetres, edge to edge, but insert the numbers 1, 2, 4, ... of the sequence obtained above (starting at the extreme left-hand edge of each strip).

2 Use these scales and Figure 4(ii) to verify that:
  a $4 \times 8 = 32$ *b* $16 \times 4 = 64$ *c* $2 \times 32 = 64$ *d* $8 \times 16 = 128$

3 Adapt Figure 3(ii) to show a scheme for dividing one number by another. Use your scales to find the quotients:
  a $8 \div 2$ *b* $64 \div 4$ *c* $32 \div 32$ *d* $128 \div 4$

Find some other products and quotients.

---

3 Calibrating the multiplication scales

The scales you have made can be used for multiplication and division involving the numbers 1, 2, 4, 8, ..., but what about numbers like 3, or 7, or 29?

Arithmetic

In Section 2 we noted the correspondence between:
the elements of the set {0, 1, 2, 3, ...} under addition, and
the elements of the set {1, 2, 4, 8, ...} under multiplication.
This mapping can be shown thus:

Numbers to be added: 0   1   2   3   4   ...
                             ↕   ↕   ↕   ↕   ↕
Numbers to be multiplied: 1   2   4   8   16 ...

Figure 6 shows the graph of the mapping of one set to the other.

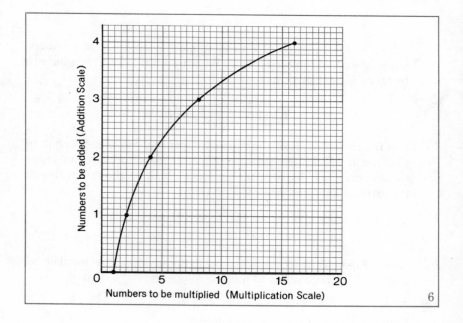

6

The points lie on a smooth curve, and it is reasonable to assume that for any number on the multiplication scale we can obtain from the graph the corresponding number on the addition scale. For example 10 on the multiplication scale corresponds to 3·3 approximately on the addition scale, so you can mark 10 on your strip, 3·3 units from the left-hand end.

The slide rule

Exercise 3B

1. Copy and complete this table, using the graph in Figure 6:

Numbers to be multiplied	1	1·5	2	2·5	3	4	5	6	7	8	9	10	11	12	13	14	15	16
Numbers to be added	0	0·6	1	1·3			2			3		3·3				3·7		4

2. Calibrate a sheet of 2-mm squared paper and 2-mm strip, taking a scale of 2 cm to represent 1 unit as shown in Figure 7. To do this use the table in question *1*, and mark 1 at the left-hand end, then mark '1·5' 0·6 units along, then mark '2' 1 unit along, and so on.

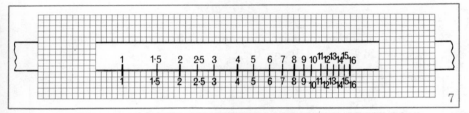

7

3. Use your squared paper and strip to calculate the products and quotients below. Remember that the positions of the numbers on your scales are approximate, and therefore your answers will be approximate also.

a	2×3	*b*	3×5	*c*	4×4	*d*	7×2	*e*	6×2·5
f	4×1·5	*g*	8×1·5	*h*	2·5×2·5	*i*	12÷4	*j*	16÷8
k	9÷3	*l*	12÷6	*m*	16÷10	*n*	10÷2·5	*o*	6÷1·5

4 The slide rule

Look at your slide rule. Notice the C and D scales, which are like the ones you have been using. Place the hairline on the *cursor* over 2 on the D scale, then move the *slide* until the 1 on the C scale is under the hairline. You will find it helpful to hold the *body* of the slide rule in such a way that you can lever the slide very slowly for the last adjustment. Place the hairline on the cursor over 3 on the C scale, and read off the number on the D scale which is under the hairline. Now place the hairline over 4 on the C scale and read off the corresponding number on the D scale.

Arithmetic

The C and D scales contain the whole numbers from 1 to 10, with the divisions between these accurately subdivided as follows:

> 1 to 2, divided into tenths, and each tenth divided into tenths;
> 2 to 4, divided into tenths, and each tenth divided into fifths.
> 4 to 10, divided into tenths, and each tenth divided into halves.

In order to give you some practice in reading numbers accurately on the scales, the letters in the following diagrams indicate numbers on the C or D scale.

Exercise 4

1 Read off, or write down, the numbers corresponding to the letters shown in Figures 8, 9 and 10.

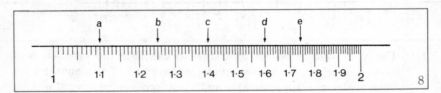

8

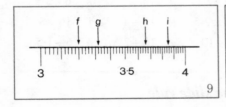

9

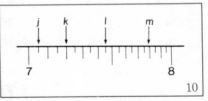

10

2 Setting the cursor first, slide the C scale to the right until the 1 is opposite each of the following numbers on the D scale in turn. In each case read off, or write down, the number on the D scale corresponding to 2 on the C scale.

    a 1·6   *b* 3·1   *c* 4·5   *d* 1·24   *e* 2·22   *f* 4·9

What kind of calculation have you been making here?

> *Note: From the above Exercise you will notice that 3-figure accuracy is the best you can expect to obtain when using a slide rule.*

5 Multiplication and division

Look at the numbers 1, 2, 3, 4,..., 10 on each of the C and D scales on a slide rule. From Figure 11(i) we can find the product 2×3.5 as in Figure 11(ii) by moving the slide (and scale C) to the right until 1 on scale C is opposite 2 on scale D, then looking along scale C to 3·5 and reading off the product on scale D.

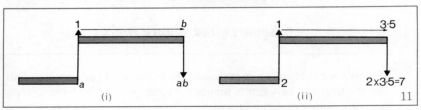

When the slide rule is used in this way, the first and final numbers are on scale D.

It is important to remember that all the slide rule is giving is the *sum of two lengths*, one on the C scale and the other on the D scale; that these lengths represent numbers, as on the number line; and that the rule is calibrated in such a way that the *product* of the two numbers results.

Exercise 5

Use a slide rule to calculate the following products:

1	2×3	*2*	2×5	*3*	3×2.5	*4*	3×1.2
5	4×2.5	*6*	4×1.6	*7*	5×2	*8*	5×1.8
9	1.4×2.5	*10*	1.8×5	*11*	2.2^2	*12*	3.14^2

From Figure 12(i) we can find the quotient $6.8 \div 2$ as in Figure 12(ii). Note that here the slide rule gives us the *difference of two lengths*.

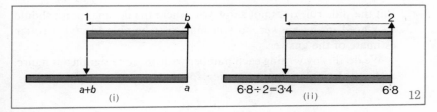

Arithmetic

Exercise 6

Use a slide rule to calculate the following quotients:

1   10 ÷ 5   *2*   8 ÷ 4   *3*   7 ÷ 2   *4*   10 ÷ 4

5   10 ÷ 2·5   *6*   8 ÷ 6   *7*   8 ÷ 7   *8*   4 ÷ 2·5

9   5 ÷ 3   *10*   3·5 ÷ 2·5   *11*   3·14 ÷ 2   *12*   3·14 ÷ 3

6 Squares and square roots

If your slide rule has an A scale, use the hairline on the cursor to find the relationship between a number on the A scale and the corresponding number on the D scale.

Exercise 7

Use the A and D scales to write down the square roots of:

1   30   *2*   66   *3*   3   *4*   1·54   *5*   20   *6*   88

Use the A and D scales to write down the squares of:

7   9   *8*   3·6   *9*   5·1   *10*   7·2   *11*   1·9   *12*   1·08

7 Rough estimates

Now that you know how to use a slide rule, you are ready to work out harder calculations such as

28×13; $594 \div 33$; 0.87×0.52; 0.063×0.0088; $\dfrac{365 \times 3.14 \times 8.5}{6.25 \times 60}$

But the slide rule does not show you where the decimal point should be placed in the answer, so you must first be able to give a rough estimate of the answer.

We do this by writing each number with only one significant figure, and calculating an approximation for the answer.

Rough estimates

Example 1. 28×13
$\doteqdot 30 \times 10$
$= 300$

Example 2. $\dfrac{594}{33}$
$\doteqdot \dfrac{600}{30}$
$= 20$

Example 3. 0.87×0.52
$\doteqdot 0.9 \times 0.5$
$= 0.45$
$\doteqdot 0.4$

Example 5. $\dfrac{365 \times 3.14 \times 8.5}{6.25 \times 60}$
$\doteqdot \dfrac{400 \times 3 \times 8}{6 \times 60}$
$= \dfrac{400 \times 24}{360}$
$\doteqdot \dfrac{400 \times 20}{400}$
$= 20$

Example 4. $0.063 \div 0.0088$
$\doteqdot \dfrac{0.06}{0.009}$
$= \dfrac{60}{9}$
$\doteqdot 7$

Exercise 8

Give rough estimates of the answers to questions *1-14*. Your estimates may not agree with those given in the Answers, but the positions of the decimal points should be the same.

1 37×29
2 38.9×2.4
3 31.7×0.48

4 8.7×0.07
5 0.0536×0.429
6 3.14×281

7 $289 \div 48$
8 $83.7 \div 5.8$
9 $6.37 \div 248$

10 $0.48 \div 6.9$
11 $\dfrac{2.89 \times 760}{46.3}$
12 $\dfrac{3.14 \times 1040}{29 \times 0.68}$

13 5×8.9^2
14 $67 \times 76 \times 0.94$

15 Estimate the area of a rectangle 7·56 centimetres long and 2·14 centimetres broad.

16 Estimate the area of a square of side 0·23 centimetre.

17 Estimate the volume of a cuboid 3·67 m long, 2·33 m broad, and 0·875 m high.

Arithmetic

18 Estimate the perimeter of a plot of ground with straight boundary walls of lengths 180 m, 135 m, 256 m, 93 m, and 106 m.

19 1 cm³ of gold has a mass of 19·3 g, and 1 cm³ of iron has a mass of 7·86 g. Estimate the mass of:

a 428 cm³ of gold   *b* 2750 cm³ of iron

20 The circumference of a planet is 6·28 times its radius. Taking the radius of the earth to be 6340 km, estimate its circumference. Taking the moon's radius to be 1730 km, estimate its circumference.

The rough estimate is an essential part of every slide rule calculation, and should always be written down or worked out mentally.

Exercise 9

In each of the following examples, find a rough approximation, and then calculate the product using a slide rule:

1	11 × 19	*2*	13 × 18	*3*	21 × 18
4	3·2 × 25	*5*	4·1 × 190	*6*	3·7 × 2·1
7	1·16 × 5·63	*8*	1·08 × 6·09	*9*	7·14 × 106
10	421 × 0·219	*11*	1·2 × 0·78	*12*	123 × 456
13	2·15 × 0·375	*14*	0·65 × 0·145	*15*	0·523 × 0·0187

Exercise 10

Again find rough estimates, and then calculate the quotient in the following:

1	190 ÷ 13	*2*	380 ÷ 18	*3*	594 ÷ 33
4	7·8 ÷ 2·3	*5*	97 ÷ 4·8	*6*	385 ÷ 204
7	79·4 ÷ 3·5	*8*	824 ÷ 41	*9*	3·87 ÷ 15·8
10	0·43 ÷ 0·19	*11*	0·007 ÷ 6·2	*12*	0·86 ÷ 0·029

If you try to multiply 7 by 5 you will finish up off the scale. Since you are taking care of the decimal point separately, it is possible to set the 10 opposite 7 instead of 1, and then to read off the product opposite 5 on the D scale.

Successive multiplication and division

This is shown in the second diagram of Figure 13(i), where a is set opposite 10 and the product is read opposite b.

In division, the quotient will be found opposite either 1 or 10 on the C scale, as indicated in Figure 13(ii).

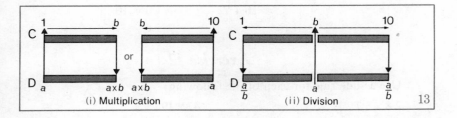

Exercise 11

Work out the following by slide rule:

1	7×5	2	78×83	3	$4 \cdot 9 \times 9 \cdot 6$
4	$0 \cdot 83 \times 0 \cdot 75$	5	$2 \cdot 76 \times 94 \cdot 7$	6	$63 \div 8$
7	$0 \cdot 92 \times 4 \cdot 8$	8	$0 \cdot 64 \times 0 \cdot 185$	9	$0 \cdot 059 \times 0 \cdot 895$
10	$3 \div 6$	11	$100 \div 33 \cdot 3$	12	$0 \cdot 87 \div 0 \cdot 092$
13	$2 \cdot 8 \div 79 \cdot 3$	14	$0 \cdot 725 \div 9 \cdot 6$	15	$0 \cdot 046 \div 0 \cdot 725$
16	$0 \cdot 24 \div 6 \cdot 28$	17	$462 \div 0 \cdot 518$	18	$0 \cdot 832 \div 94 \cdot 3$

8 Successive multiplication and division

When you have to perform successive multiplication and/or division, each calculation is done in turn; and since the product or quotient is always on the D scale, it is set ready to act as the a in the next calculation. If you have both multiplication and division to do, take them alternately, starting with division. Try it, and you will see why. You may discover short cuts; for example, a calculation which is very common in proportion can be done as follows:

$$\frac{7 \cdot 54 \times 3 \cdot 26}{5 \cdot 93} \qquad 3 \cdot 26 \leftarrow\!\!\text{———}\!\!5 \cdot 93$$
$$\downarrow \qquad\qquad \uparrow$$
$$\text{Answer} \qquad 7 \cdot 54$$

Arithmetic

Note how easy it is to use a slide rule to calculate a percentage.

Example. Express 43 as a percentage of 262.

$$\text{Percentage} = \frac{43}{262} \times 100 \qquad \frac{40 \times 100}{300}$$
$$= 16 \cdot 4 \qquad \qquad \div 10$$

Exercise 12

Use a slide rule for each of the following:

1. $\dfrac{2 \cdot 8 \times 3 \cdot 7}{1 \cdot 9}$
2. $\dfrac{47 \times 19}{73}$
3. Calculate the area of a rectangle 3·78 cm long and 2·79 cm broad.
4. Give the square roots of: *a* 29 *b* 1650
5. Express as percentages: *a* $\frac{4}{7}$ *b* $\frac{5}{8}$ *c* $\frac{8}{9}$ *d* $\frac{5}{32}$ *e* $\frac{9}{16}$
6. $\dfrac{5 \cdot 2 \times 6 \cdot 8}{7 \cdot 7}$
7. $\dfrac{35 \cdot 6}{16 \cdot 1 \times 3 \cdot 82}$
8. Calculate the area of a square of side 8·8 cm.
9. The area of a rectangle is 3·96 m² and its breadth is 0·92 m. Calculate its length.
10. £1 is equivalent to 2·78 dollars. How many dollars can be obtained with £686?
11. A profit of 13% is made on goods costing £789. What was the selling price (to the nearest £)?
12. A loss of 7% is made on goods costing £893. Find the selling price (to the nearest £).
13. $4 \cdot 8 \times 7 \cdot 6 \times 0 \cdot 6$
14. $\sqrt{\dfrac{5 \cdot 7}{2 \cdot 5}}$
15. Calculate the volume of a cuboid of length 1·6 m, breadth 1·4 m, and height 0·7 m.
16. A cuboid has volume 15·2 cm³, length 7·8 cm, and breadth 3·9 cm. Calculate its height.

Successive multiplication and division

17 An aircraft flies 3790 km in 6 hours 45 minutes. Calculate its average speed.

18 A car travels 393 km at an average speed of 58·5 km/h. How long does its journey take?

19 27 books weigh 6·5 kg. Find the weight of 37 similar books.

20 $6·8^2 \times 3·4^2$

21 $\dfrac{36 \times 49}{25 \times 31}$

22 For further practice, calculate the answers to Exercise 8.

Arithmetic

Summary

1 A rough estimate must always be made

e.g. $38·9 \times 0·723$ and $\dfrac{38·9}{0·723}$

$\ \ \doteqdot 40 \times 0·7 \qquad\qquad \doteqdot \dfrac{40}{0·7} = \dfrac{400}{7}$

$\ \ = 28·0$

$\ \ \doteqdot 30 \qquad\qquad\qquad \doteqdot 60$

2 Multiplication by slide rule

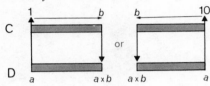

3 Division by slide rule

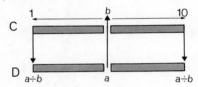

4 Squares by slide rule

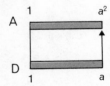

5 Square roots by slide rule

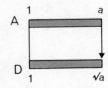

Note: On some slide rules the A scale is below the D scale.

Note to the Teacher on Chapter 3

The main aims of Chapter 3 are to indicate the relation between the lengths of the circumference and the diameter of a circle, to introduce the formulae for the circumference and area of a circle, and to apply these formulae in a variety of ways.

A practical approach is suggested both to length and to area, in order to encourage the pupils to discover and accept the various results. Proofs of the necessary formulae cannot be given at this stage, but the methods employed can be taken at a more sophisticated level later, in calculus, when the perimeter of curves and the areas bounded by curves are investigated.

The idea in *Section* 1 is to find two lengths between which the length of the circumference of a given circle must lie. These lengths are given by the perimeters of figures which are inscribed in the circle and circumscribed about the circle. It is intended that pupils do questions *1* and *2*, or *1* and *3*, of Exercise 1. Question *3* follows more logically from question *1*, but makes greater demands on the pupils' knowledge of geometry and algebra. Question *4* is well worth spending time on, and with reasonable care in measuring a remarkably good approximation for π can be obtained. It is worth emphasizing that for *every* circle the circumference is approximately three times the diameter in length.

In question *1* of *Section* 4 the effect is very striking if contrasting colours are used for the two halves of the circle.

It will be noticed that in the Worked Examples 2 in *Section* 3 and in *Section* 5, two methods are suggested; the teacher must decide which is preferable at this stage—direct substitution, or change of subject of the formula.

The chapter has been placed immediately after the chapter on the slide rule, partly because this work shows the advantages of a mechanical aid to computation and provides practice in the use of the slide rule. It should be pointed out that the approximation for π marked on most slide rules is 3·14 so that answers will be obtained to 2 significant figures; however, if π is taken as $\frac{22}{7}$ answers should be rounded off to 3 significant figures where necessary.

The Circumference and Area of a Circle

1 The circumference of a circle

Exercise 1–Practical

1. Draw a circle of radius 1 unit, and by 'stepping' the radius round the circumference inscribe a regular hexagon in the circle as shown in Figure 1.

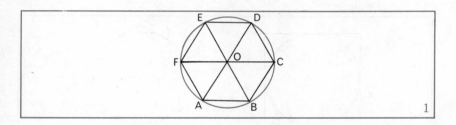

1

- a What is the length of AB?
- b What kind of triangle is OAB?
- c What is the perimeter of the hexagon?
- d What is the length of the diameter of the circle?
- e Is the length of the circumference greater than, or less than, the perimeter of the hexagon?
- f What is the value of the ratio $\frac{\text{perimeter of the inscribed hexagon}}{\text{length of the diameter of the circle}}$?

From the answers to the above questions it should be seen that *the length of the circumference of a circle is more than 3 times the length of its diameter.*

Arithmetic

2 Draw a square of side 2 units. You are now going to inscribe a circle in this square, i.e. draw a circle which fits exactly inside the square. Where must the centre of this circle lie? See Figure 2(i).

- **a** What is the length of the radius of the circle?
- **b** What is the length of the diameter of the circle?
- **c** What is the perimeter of the square?
- **d** Which is greater, the perimeter of the square or the length of the circumference of the circle?
- **e** What is the value of the ratio $\dfrac{\text{perimeter of the square}}{\text{length of the diameter of the circle}}$?

From the answers to the above questions, it should be seen that *the length of the circumference of a circle is less than 4 times the length of its diameter.*

From questions *1* and *2* it is clear that the length of the circumference of a circle is between 3 and 4 times the length of the diameter of the circle.

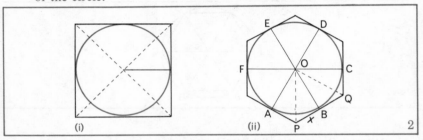

3 (*This may be taken as an alternative to question 2.*)

Draw a circle as in question *1*, but this time draw lines perpendicular to the radii at A, B, C, D, E, F to form a hexagon circumscribing the circle as shown in Figure 2(ii).

- **a** Is the length of the circumference of the circle greater than, or less than, the perimeter of this hexagon?
- **b** What are the sizes of angles AOB, POB, BOQ, POQ?
- **c** What kind of triangle is triangle OPQ?
- **d** If the length of PB is x units, what are the lengths of PQ and OP?

Using Pythagoras' theorem, $(2x)^2 = x^2 + 1^2$
$\Leftrightarrow \quad 4x^2 = x^2 + 1$
$\Leftrightarrow \quad 3x^2 = 1$
$\Leftrightarrow \quad x^2 \doteqdot 0\cdot333$
$\Leftrightarrow \quad x \doteqdot 0\cdot58$, since x represents a positive number.

The circumference of a circle

e Can you now give the perimeter of the hexagon?

f What is the value of the ratio $\dfrac{\text{perimeter of the circumscribed hexagon}}{\text{length of the diameter of the circle}}$?

From the answers to the above questions, *the length of the circumference of a circle is less than 3·48 times the length of its diameter.*

From questions *1* and *3* it is clear that the length of the circumference of a circle is between 3 and 3·48 times the length of the diameter of the circle.

4 *a* Here you will require several circular or cylindrical objects, e.g. bicycle wheel, tin, chalk circle on the floor, etc. Devise methods of measuring the diameter and the circumference of each of these (using tape measure, string, shoes, etc.). Copy and complete this table:

Length of diameter	Length of circumference	$\dfrac{\text{Length of circumference}}{\text{Length of diameter}}$

b Calculate your average for the ratio $\dfrac{\text{length of circumference}}{\text{length of diameter}}$, and then find the average for the whole class. (It is unlikely that the third significant figure will be dependable.)

Formulae for the length of the circumference

The approximate results above suggest that the ratio $\dfrac{\text{length of circumference}}{\text{length of diameter}}$ is the same for all circles. This is in fact so, and the ratio is called π (the Greek letter p, pronounced 'pie'), so we have

$$\frac{C}{d} = \pi,$$

where C represents the length of the circumference and d represents the length of the diameter in the same units.

It follows that, if in the same units
C represents the length of the circumference,
d represents the length of the diameter,
r represents the length of the radius,
then $C = \pi d$, and $C = 2\pi r$.

Arithmetic

2 Approximations for π

The number π cannot be expressed exactly as a common or decimal fraction; it is an *irrational number*, and is placed on the number line between 3·141 and 3·142.
  Approximations for π are:
  3·14   rounded off to 3 significant figures
  3·142   rounded off to 4 significant figures
  3·1416 rounded off to 5 significant figures

  The fraction $\frac{22}{7}$ when expressed as a decimal is equal to 3·142857...; rounded off to 3 significant figures this is 3·14, and to 4 significant figures it is 3·143. Therefore π may be taken to be $\frac{22}{7}$ *when we require to work only to 3 significant figures.* But remember that when π is taken to be $\frac{22}{7}$, we must not give more than 3 significant figures in any result obtained.

Historical note

For many centuries—probably since man invented the wheel, and realized that a 'tyre' was desirable to prevent wear—interest has been shown in this number which we now call π. In the Bible, in I Kings vii.23, you will find that the Jews of King Solomon's time apparently took π to be 3; you may find in the school library books which give the story of how better approximations for π were obtained. By using electronic computers it was actually worked out in 1961 to 100 265 decimal places!

3 Calculations involving circumferences of circles

Where it is clear that by *circumference* we mean *length of circumference*, etc., we shall use this shorter form.
  The circumference of a circle is given by $C = \pi d$ or $C = 2\pi r$.
  The use of a slide rule is helpful. Some slide rules have the position of π marked on each scale.

Calculations involving circumferences of circles

Example 1. Calculate the circumference of a wheel which has a diameter of 23 cm.

In $C = \pi d$, put $d = 23$
Then $C = 3 \cdot 14 \times 23$
$= 72 \cdot 2$

The circumference is approximately 72 cm.

Example 2. A wheel has a radius of 28 cm. How many times will it turn on a vehicle that travels 440 m?

(You can see, by rolling a wheel along the floor, or a coin along the desk, that the distance travelled in one complete turn is equal to the circumference.)

In $C = 2\pi r$, put $r = 28$
Then $C = 2 \times \frac{22}{7} \times 28$
$= 176$

i.e. the distance travelled during 1 complete turn = 176 cm.
The total distance travelled = 440 m
$= 44000$ cm
The number of turns = $\frac{44000}{176}$
$= 250$ approximately.

Example 3. The distance round the edge of a circular pond is 60 metres. Find the radius of the pond.

First method
In $C = 2\pi r$, put $C = 60$.

$60 = 2 \times 3 \cdot 14 \times r$

$\Leftrightarrow 60 = 6 \cdot 28 r$

$\Leftrightarrow r = \dfrac{60}{6 \cdot 28}$

$= 9 \cdot 55$

Second method
$C = 2\pi r$

$\Leftrightarrow r = \dfrac{C}{2\pi}$

Replacing C by 60,

$r = \dfrac{60}{2 \times 3 \cdot 14}$

$= 9 \cdot 55$

The radius is approximately 9·55 metres.

Exercise 2

Use $3 \cdot 14$ or $\frac{22}{7}$ as an approximation for π.

1 Find the circumferences of circles with diameters of lengths:

a 7 cm   *b* 21 cm   *c* 35 cm   *d* 49 cm
e 10 m   *f* 4 cm   *g* 8 mm   *h* 2·4 m

Arithmetic

2. Find the circumferences of circles with radii of lengths:
   - a 14 cm
   - b 21 cm
   - c 28 cm
   - d 56 cm
   - e 2 m
   - f 10 m
   - g 5 m
   - h 8·1 m

3. The diameter of a penny is 20 mm. Calculate its circumference.

4. The diameter of a record player turntable is 28 cm. Calculate its circumference.

5. The radius of a record is 15 cm. Calculate its circumference.

6. The length of the minute hand of a watch is 1 cm. What distance does the point of the hand travel in: a 1 hour   b 12 hours?

7. The centre circle of a football pitch has a radius of 3 metres. What is the length of the white line forming this circle?

8. The semicircle on a hockey pitch has a radius of 6·3 metres. What is the length of the white line forming this semicircle?

9. A boy flies a model aircraft at the end of a horizontal wire 20 metres long. What distance does the aircraft travel in flying once round the circle?

10. Calculate to two significant figures the circumference of the earth, assuming that it is a sphere with radius 6400 kilometres.

11. The diameter of a motor car's wheel is 42 cm.
    - a Calculate the circumference of the wheel.
    - b How far, in metres, does the car travel while the wheel turns 50 times?

12. Figure 4 shows the plan of a football pitch, with semicircular areas behind the goals. If the pitch is a rectangle 100 m long and 70 m broad, calculate the length of one lap of the track surrounding the pitch.

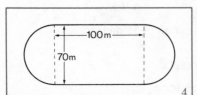
4

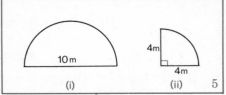
(i)   (ii)   5

13. Calculate the perimeters of the shapes in Figure 5. (i) shows a semicircle and diameter; (ii) shows a quarter circle and two radii.

Calculations involving circumferences of circles

14 Calculate the diameters of circles with circumferences of:
   a 44 cm   *b* 55 m   *c* 110 mm   *d* 15 m

15 The circumference of a circular plate is 66 cm. Calculate:
   a its diameter   *b* its radius.

16 A sports master wants to mark out a circular 400 metre running track. What radius should he use?

17 The roundabout at a road junction has to have a minimum circumference of 110 metres. What is the minimum diameter of the roundabout?

18 It is found, by using a tape measure, that the circumference of a hot water tank is 2 metres. What is the diameter of the tank (to the nearest centimetre)?

19 The turning circle of a car is 60 metres long. What is the radius of the turning circle?

20 The circumference of a circular paper-weight at its widest part is 20 cm. Calculate the radius at that part.

Exercise 2B

1 Calculate the circumferences of circles with:
   a diameters of (*1*) 63 mm (*2*) 3·5 cm (*3*) 100 m
   b radii of   (*1*) 84 mm (*2*) 10·5 cm (*3*) 3 km

2 A ship sails a complete circle with a radius of 1750 metres. What is the circumference of the circle in kilometres?

3 Two concentric circles (i.e. having the same centre) have diameters of 30 metres and 51 metres. What is the difference in the lengths of their circumferences?

4 In Figure 6, quarter circles have been cut from the corners of a rectangular sheet of metal. Calculate the perimeter of the resulting shape.

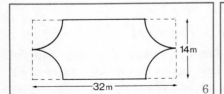

6

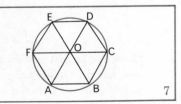

7

Arithmetic

5   Figure 7 shows a regular hexagon inscribed in a circle.
a What is the size of angle AOB, and what fraction is this of one complete turn?
b What fraction of the circumference is arc AB?
c If the radius of the circle is 1 cm, calculate the length of arc AB.

6 *a* Find the circumference of a circle with radius 14 cm.
  b Calculate the lengths of arcs of this circle which are cut off by drawing at the centre angles of: (*1*) 90° (*2*) 45° (*3*) 120° (*4*) 270°.

7   A wheel has a radius of 28 cm. Calculate its circumference, and the number of turns it will make on a vehicle which travels 440 m.

8   Calculate the radii of circular racing tracks with circumferences:
a 157 m   *b* 220 m   *c* 628 m   *d* 1500 m

9   The wheel of a car turns 80 times while the car travels 400 metres.
a Calculate the circumference of the wheel.
b Calculate the radius of the wheel, in centimetres.

10 An artificial satellite travels in a circular orbit 1300 km above the earth's surface. If it completes each orbit in 2 hours, and the radius of the earth is 6400 km, calculate the speed of the satellite in km/h.

11 2000 years ago a Greek astronomer measured a distance of 800 km on the earth's surface, which he estimated was $\frac{1}{48}$ of the circumference. What value did this give him for the radius of the earth (to two significant figures)?

12 Calculate the distance travelled by the earth round the sun in one year, assuming that it moves in a circle of radius $1 \cdot 50 \times 10^8$ km. Give your answer in standard form ($a \times 10^n$), with a rounded off to two significant figures.

4 The area of a circle

Exercise 3–Practical

1   Draw the largest circle you can inside a square of gummed paper of side 10 cm. Divide this circle into sectors by drawing angles of 30° at the centre as shown in Figure 8(i). Bisect one of these sectors.

The area of a circle

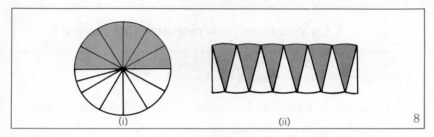

8

Now cut out all the sectors and stick them side by side as in Figure 8(ii). Approximately what shape do you get? Can you suggest how this shape could be made even more nearly a rectangle? If the radius of the circle is r, what is the breadth of the 'rectangle'? What is the length of the 'rectangle'? What is the area of the 'rectangle'?

It will be seen that, by taking a very large number of sectors of the circle, the shape in Figure 8(ii) will become indistinguishable from a rectangle, and that its breadth is r units and its length is $\frac{1}{2}C$ units, i.e. $\frac{1}{2} \times 2\pi r = \pi r$ units. So the area A square units of the circle is given by $r \times \pi r$ square units, i.e.

$$A = \pi r^2$$

Note.—Since $r = \frac{1}{2}d$, $r^2 = \frac{1}{2}d \times \frac{1}{2}d = \frac{1}{4}d^2$, the formula for the area of a circle can be written $A = \frac{1}{4}\pi d^2$; this form is often used by engineers since with callipers they find it easier to measure the diameter of a circle than the radius.

2 Draw a circle of radius 5 cm on 2-mm squared paper. By counting squares in a quarter circle find an estimate of the area of the circle. (You should consider what allowance to make for 'broken' squares.) Verify that this confirms the result $A = \pi r^2$. Repeat for circles of various radii.

3 Use the values of A and r obtained in question *2* to draw a graph to illustrate the relationship between A and r^2.

4 Use the values of C and r obtained in Section 1, question *4*, to draw a graph to illustrate the relationship between C and r.

5 What do you notice about the shape of the graphs in questions *3* and *4*?
Notice that since $A = \pi r^2$, A is directly proportional to r^2. What is the relationship between C and r?

Arithmetic

5 Calculations involving areas of circles

The area of a circle is given by $A = \pi r^2$.

Example 1. Calculate the area of a circle with diameter 7 cm.

$d = 7$ $\quad\quad A = \pi r^2 \quad\quad$ or $\quad\quad A = \pi r^2$
$r = \frac{7}{2} \quad\quad\quad\; = \frac{22}{7} \times \frac{7}{2} \times \frac{7}{2} \quad\quad\quad\quad\; = 3{\cdot}14 \times 3{\cdot}5^2$
$\quad\quad\quad\quad\quad\; = \frac{77}{2} \quad\quad\quad\quad\quad\quad\quad\; = 38{\cdot}46$
$\quad\quad\quad\quad\quad\; = 38{\cdot}5$

The area is approximately 38·5 cm².

Example 2. The area of a circular pond is 67 m². Find the diameter of the pond.

First method $\quad\quad\quad\quad\quad\quad\quad\quad$ *Second method*

In $A = \pi r^2$, put $A = 67$ $\quad\quad\quad\quad A = \pi r^2$
$67 = 3{\cdot}14 r^2$ $\quad\quad\quad\quad\quad\quad \Leftrightarrow r^2 = \dfrac{A}{\pi}$
$\Leftrightarrow r^2 = \dfrac{67}{3{\cdot}14} = 21{\cdot}3$
$\Leftrightarrow r = \sqrt{21{\cdot}3} \quad$ (r is positive) $\quad \Leftrightarrow r = \sqrt{\dfrac{A}{\pi}} \quad$ (r positive)
$\quad\quad\; = 4{\cdot}62 \quad$ (from tables) $\quad\quad\;$ Replacing A by 67,
$\quad\quad\quad\quad\quad\quad\quad\quad\quad\quad\quad\quad\quad\quad r = \sqrt{\dfrac{67}{3{\cdot}14}}$
$\quad\quad\quad\quad\quad\quad\quad\quad\quad\quad\quad\quad\quad\quad\;\; = \sqrt{21{\cdot}3}$
$\quad\quad\quad\quad\quad\quad\quad\quad\quad\quad\quad\quad\quad\quad\;\; = 4{\cdot}62$

The radius is approximately 4·62 m, so the diameter is about 9·2 m.

Exercise 4

Use 3·14 or $\frac{22}{7}$ as an approximation for π.

1 Find the areas of circles with radii of lengths:

    a   7 cm   *b*   14 cm   *c*   10 cm   *d*   2 cm

2 Find the areas of circles with diameters of lengths:

    a   7 mm   *b*   2 cm   *c*   10 m   *d*   1 km

Calculations involving areas of circles

3 The diameter of a penny is 20 mm. Calculate the area of one of its flat surfaces.

4 Calculate the area of a circular mirror of radius 21 cm.

5 The length of the minute hand of a watch is 1 cm. What area does the hand sweep over in 1 hour?

6 The semicircle on a hockey pitch has a radius of 6·3 metres. What is the area of this semicircle?

7 A long-playing record has a diameter of 30 cm. Calculate the area of one side of the record.

8 Calculate the area of the sports pitch, consisting of a rectangular area, with semicircular ends, shown in Figure 9.

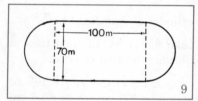
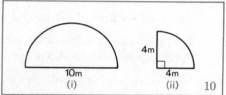

9 Calculate the areas of the shapes shown in Figure 10; they consist of a semicircle, and a quarter circle, respectively.

10 Calculate the area of each of the shaded regions in Figure 11. In (i), the inner boundary is a circle and the outer boundary is a square; in (ii), the region is bounded by a semicircle and three sides of a rectangle.

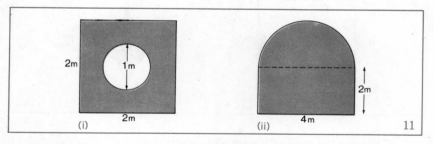

11 Calculate the radii of circles with areas:

   a 314 cm^2 *b* 154 cm^2 *c* 22 cm^2 *d* 123 cm^2

Arithmetic

12 A circular sheet of metal has an area of 88 m². Calculate its diameter.

13 The area of a circular plot of ground is 616 m². Calculate its radius.

14 The area of a circular window has to be at least 75 m². Calculate the minimum possible radius of the window.

15 The area of a circular sheet of metal is 1250 cm². Calculate:
     a the radius of the plate   *b* the circumference of the sheet.

Exercise 4B

1 Calculate the areas of circles with:
     a radii of   (*1*) 3·5 cm   (*2*) 100 m
     b diameters of (*1*) 3·5 cm   (*2*) 8 m

2 Calculate the area of the base of a cylinder of diameter 12 cm.

3 A square lawn of side 6 m has a circular flower bed of diameter 4 m in the centre. Make a sketch, and calculate the area of grass.

4 Milk-bottle tops are made by cutting as many circles as possible of diameter 6 cm out of a strip of aluminium 360 cm long and 6 cm broad. What percentage of the aluminium is wasted?

5 Calculate the area of each of the shaded regions in Figure 12. In (i), the region is bounded by a quarter circle and two sides of a square; in (ii) the region is bounded by four quarter circles.

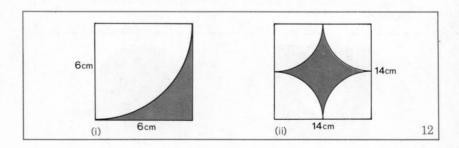

12

6 Calculate the radii of circles with areas:
     a 44 mm²   *b* 484 cm²   *c* 1000 m²

7 Find the diameter of a circular ink blot which covers an area of 2 cm².

Calculations involving areas of circles

8   Calculate the circumference of a circular metal plate with area 15·7 m².

9   The ratio of the radii of two circles is 2:1. What is the ratio of:
a   their circumferences   *b*   their areas?

10   Examine a solid cylinder. How many flat surfaces does it have? How many curved surfaces? Wrap a piece of paper round the curved surface and cut it so that it exactly fits the surface. Now unwrap the paper and lay it out flat. What shape is it?

What is the connection between the *length* of the paper and the circle forming the end of the cylinder? If the radius of the cylinder is 7 cm, what must be the *length* of the paper? If the height of the cylinder is 5 cm, what must be the *breadth* of the paper? Can you now calculate the area of the curved surface of the cylinder?

By this method, calculate the area of the curved surface of each of the following cylinders:

a   radius of the end = 2 cm, height = 10 cm
b   radius of the end = 3 cm, height = 6 cm
c   radius of the end = 2·8 cm, height = 4 cm.

An approximation for π

Mark off on a large sheet of paper a set of parallel lines 4 cm apart. Cut a large number of sticks (e.g. matchsticks) each 2 cm long. Now place the sheet of paper on a flat surface and drop the matches one by one on to it from about a metre above it. Count the number of sticks which cross or touch the parallel lines. Tabulate your results thus:

Number of sticks dropped	Number of sticks touching or crossing a line

Evaluate the ratio $\dfrac{\text{number of sticks dropped}}{\text{number touching or crossing a line}}$

Your answer should be a good approximation for π!

Arithmetic

Summary

1. The value of the ratio $\frac{\text{circumference}}{\text{diameter}}$ is the same for all circles and is denoted by π.
$$\frac{C}{d} = \pi$$

2. *Approximations for π are:*
   3·14, to 3 significant figures, or $\frac{22}{7}$

3. *Circumference of a circle* : $C = 2\pi r$, or $C = \pi d$

4. *Area of a circle* : $A = \pi r^2$

5. When 3·14 or $\frac{22}{7}$ are used as approximations for π, answers should not be given to more than 3 significant figures.

Note to the Teacher on Chapter 4

The aims of Chapter 4 are to familiarize pupils with some of the vocabulary of statistics:
population, sample, class interval, frequency, histogram, frequency polygon, mean, median, mode;
and to illustrate the use of statistics in a variety of situations.

Section 1 contains some revision of earlier work, and at the same time introduces the idea of a statistical population and of samples from that population. This is related to previous ideas in the Chapter on Probability, and several questions in the present chapter refer to the relative frequency of an outcome of a trial in some sample, and the probability of the outcome in the associated population.

The organization and illustration of data are developed by means of frequency tables in *Sections* 2 and 7, and histograms and frequency polygons in *Sections* 4 and 5 respectively. The tally system for constructing a frequency table is worth practising as it is easy, accurate, and has widespread uses.

Section 3 brings out the difficulty that often arises in choosing a good class interval. The choice can be improved with experience, but remains arbitrary and involves a compromise between competing requirements—namely, to reduce labour and yet to give an adequate representation of the distribution.

In *Section* 4 it should be emphasized that in a histogram the area of each rectangle is proportional to the corresponding frequency. The difficulties of boundaries which arise particularly with continuous variates (e.g. height, weight) have not been discussed in this chapter; rather are they avoided at this stage.

Section 6 attempts to emphasize that the mean, median and mode are all indicators of the 'central tendency' of a distribution, as illustrated in Figure 10. A difficulty arises in that 'average' is widely used in a colloquial sense, when mean is really intended. A suitable stress has been put on the various terms that are in common use. Throughout, calculations have been kept fairly simple in order that pupils can have plenty of practice at the techniques involved.

Section 7 includes the calculation of the mean from a frequency table. It will be noticed that mid-values of class intervals have not been introduced here; this extension will be made in Book 6.

At suitable stages pupils are encouraged to carry out their own investigations in order to obtain and analyse their own data; it is hoped that time can be found for these worth-while activities.

Statistics – 2

1 Measurements, samples and populations

In Book 2 we saw that information could be illustrated and its meaning quickly grasped from pictographs, bar charts, pie charts and line graphs.

The set of items which have the characteristic in which we are interested is called a *population*, and the subset consisting of these items actually measured is called a *sample* of that population.

For example, if we are interested in the ages last birthday of pupils in their second year of secondary education, the population is the set of all such pupils and the sample is the subset consisting of the second-year pupils whose ages we actually know.

Exercise 1 (mainly for revision)

1. The line graphs in Figure 1 indicate the primary and secondary school populations in Scotland in each year from 1952 until 1962.

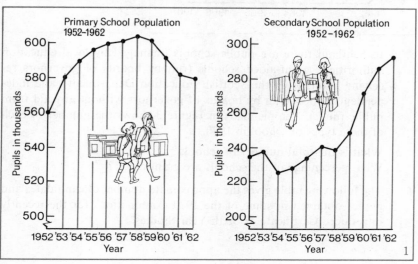

Arithmetic

- *a* The first graph shows the 'bulge' in the primary school population. In which year did it reach its peak? What relation is there to the bulge in the second graph?
- *b* In which year was the secondary school population least? Estimate the population in this year.
- *c* Calculate approximately the percentage increase in the secondary school population between 1954 and 1962.

2 Figure 2 is not unrelated to Figure 1. It shows the supply and demand for teachers in the years 1962-67.

Give as much general information as you can from this graph, e.g. the *trend* of the graph, the significance of the separation of the two heavy lines, etc.

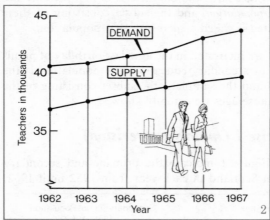

2

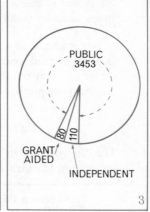

3

3 In Scotland there are public schools (managed by local education authorities), grant-aided schools (run by boards of governors receiving grants from the Scottish Education Department), and independent schools (run by boards of governors without any aid from public funds). The pie chart in Figure 3 shows the number of each of these types of school in 1970.

- *a* What is the total number of schools?
- *b* What percentage of the schools are grant-aided?

4 The following table gives the approximate number of candidates and grades obtained in some of the 1970 examinations for the recently introduced Certificate of Sixth Year Studies.

Frequency tables

Subject	Number of candidates	Grades awarded				
		A	B	C	D	E
English	1260	75	425	505	240	15
Mathematics I	360	40	90	140	70	20
Art	180	25	35	60	35	25

   a Describe the population and the samples from that population in this question.
   b Illustrate the grades for Art in a bar graph.
   c Illustrate the grades for Mathematics I in a pie chart.

5 Your own class represents a sample of the school population in the country. Suggest several features of this sample that could be measured, thereby providing information about the population.

2 Frequency tables

The marks in a class test were called out in alphabetical order of the pupils' names as follows:

```
7 6 4 6 8 3 5 5 6 1 4 4 7
7 5 9 5 4 6 3 6 2 4 5 3 5
6 5 6 4 5 5 6 3 4 5 8 7 7
```

Notice that the mark 8 occurs twice, 7 occurs five times, 6 occurs eight times, and so on.

The information can be shown more clearly in a *frequency table*:

Mark		Frequency
1	/	1
2	/	1
3	////	4
4	++++ //	7
5	++++ ++++	10
6	++++ ///	8
7	++++	5
8	//	2
9	/	1
10		0
Total		39

Arithmetic

As you can see, the original information has been transferred to the table as 'tally marks'. It is convenient in practice to use every fifth tally mark in a given score to 'cross out' the previous four, thus making it easy to count up the tally marks for each score in groups of five. The total number of tally marks for each score is called the *frequency* of the score and is entered in the frequency column. It should be noted that the tally marks give a rough picture of the distribution of marks.

The score that occurs most often is called the *mode* or *modal score* of the distribution. In the above sample, the mode is 5.

Exercise 2

1 The football scores for the English League Division I on 16th October 1971 were:

Chelsea	1	Arsenal	2	Nottingham	2	Liverpool	3
Everton	1	Ipswich Town	1	Southampton	3	Sheffield	2
Leeds Utd	3	Manchester City	0	Stoke City	1	Coventry	0
Leicester	2	Huddersfield	0	Tottenham	4	Wolverhampton	1
Manchester Utd	1	Derby County	0	West Bromwich	0	West Ham	0
Newcastle	1	Crystal Palace	2				

Construct a frequency table of the goals scored by the teams. What is the modal number of goals scored?

2 The classes in four primary schools contained the following numbers of pupils:

```
35  38  40  29  30  34  42  37  35  36  31  37
28  30  36  34  33  39  30  36  39  37  30  38
38  39  37  35  33  31  30  34  37  40  38  37
28  31  32  37  33  38  37  38  39  38  37  30
```

Make a frequency table. What is the mode?

3 The scores in the final round of the Dunlop Masters Golf Tournament in 1971 were:

```
66  70  69  68  70  72  69  71  72  74
66  63  69  69  73  73  73  69  73
69  66  67  67  70  71  69  74  80
```

Construct a frequency table. What was the modal score?

4 Construct a frequency table showing the number of children in the families of members of your class.

5 Construct a frequency table showing the number of pupils in the class born in each month of the year.

Class intervals

6 Construct a frequency table for the times (to the nearest half hour) at which pupils in your class went to bed last night. What is the mode?

7 Select at random a page from a book. Count the number of letters in each of the first hundred words on the page, recording the results in a frequency table. What is the modal number of letters per word in the sample?

3 Class intervals

When we have to consider a large number of different measurements, it is usual to group the information in *class intervals*.

Here are some examination marks:

```
42  73  54  58  85  52  48  54  60  54
58  48  70  52  53  53  53  60  25  55
60  55  50  53  75  58  52  45  65  68
58  57  82  30  55  49  57  63  72  28
```

We might decide to group the marks in class intervals of 5, starting at 25. This means that the first group would include 25, 26, 27, 28, 29, and the second group 30, 31, 32, 33, 34, and so on.

The frequency table is:

Marks		Frequency
25-29	//	2
30-34	/	1
35-39		0
40-44	/	1
45-49	////	4
50-54	//// //// /	11
55-59	//// ////	9
60-64	////	4
65-69	//	2
70-74	///	3
75-79	/	1
80-84	/	1
85-89	/	1
	Total	40

Arithmetic

Since the data, or information, have been classed together we can note, not the mode, but the *modal class*, i.e. the class interval in which the greatest frequency of marks occurs. In the above example the modal class would be 50-54 marks.

Exercise 3

1. Make a frequency table for the following marks in a mathematics examination, grouping the marks in class intervals of 5, starting at 40.

44	54	85	92	73	57	99	91	96	74	75	70
83	49	57	52	64	67	73	82	90	70	89	91
52	64	73	82	59	50	65	79	82	89	53	52

2. Make a frequency table for the data given in question *2* of Exercise 2, grouping the number of pupils in class intervals of 3, starting at 28. What is the modal class?

3. Make a frequency table for the data given in question *3* of Exercise 2, grouping the scores in class intervals of 2, starting at 63. What is the modal class?

4. Make frequency tables for the 40 examination marks given at the beginning of this Section, using class intervals of 1, 3, 10, and 20. (Perhaps you could share this work with your neighbour.) By comparing these tables decide whether you think 5 was a good choice of class interval. What is the disadvantage of

   a too small a class interval   b too large a class interval?

5. The maximum temperatures in degrees Celsius recorded at various places in Europe on 4th October 1971 were:

15	23	23	14	22	20	20	13	23	8	18	24
15	18	16	21	16	14	16	15	10	23	13	22
12	16	22	13	24	18	15	24	15	16	11	19
13	18	7	20	13	19	25	20	16	27	18	13
17	16	24	25	23	15	20	4	11	20	20	21

   Make a frequency table, using a suitable class interval.
   Compare your table with your neighbour's, and discuss how to choose a good class interval and starting point.

4 Histograms

Frequency distributions can be illustrated by means of *histograms* in which the frequency in each class is represented by a rectangle, *the area of the rectangle being proportional to the frequency*. When equal class intervals are chosen, the height of the rectangle is proportional to the frequency.

In a number of trials, 9 coins were tossed together and the following data (illustrated by the histogram in Figure 4) were obtained.

Number of heads	0	1	2	3	4	5	6	7	8	9
Frequency	2	9	36	80	131	122	83	37	10	2

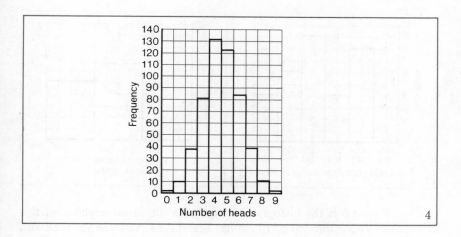

4

Note: For this sample, the *relative frequency* of '3 heads' is $\frac{80}{512} = \frac{5}{32}$.

If this is a typical sample, it follows that the *probability* of getting 3 heads when 9 coins are tossed is $\frac{5}{32}$.

* * * *

Arithmetic

So far we have been studying measurements which are whole numbers. If we now consider the height of a boy at different ages we shall be dealing with a different type of measurement, since the boy's height does not increase instantaneously from one whole number of centimetres to the next, but progressively takes up intermediate measurements.

Measurements of this type are approximations, but we can represent them conveniently by frequency tables and histograms. The table below gives the heights, to the nearest cm, of 34 boys, with the data grouped in class intervals of 5 cm.

Height in cm	140–144	145–149	150–154	155–159	160–164	165–169
Frequency	3	8	4	9	6	4

The histogram of this frequency distribution is shown in Figure 5; we see that it shows two distinct peaks.

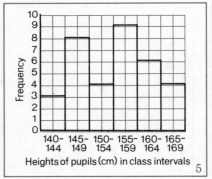

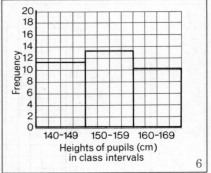

Figure 6 is the histogram drawn from the same heights, but this time a class interval of 10 cm has been taken. You can see that with too large a class interval important and interesting features of the distribution may not be apparent; with too small a class interval the regularity of a distribution may be missed.

Exercise 4

1 The number of children of school age in each household in a certain street was required for a census. Illustrate the information obtained as a histogram.

Histograms

Number of children in household	0	1	2	3	4	5	6
Frequency	4	9	44	26	7	1	1

What is the relative frequency of a household with exactly two children in this sample?

2   The weights of 40 pupils, each to the nearest kg, are given in the table. Draw a histogram of the distribution.

Weight in kg	30–34	35–39	40–44	45–49	50–54
Frequency	2	8	20	6	4

3   The following table shows the weekly wages, to the nearest £, paid to employees in a small firm. Draw a histogram of the distribution. What is the relative frequency of a wage of £15-£19 in the firm?

Wage in £	10–14	15–19	20–24	25–29	30–34	35–39	40–44
Frequency	9	7	15	24	10	4	1

4   The first-year pupils in a school were asked the number of coins they had with them, and the following information was obtained:

Number of coins	0	1	2	3	4	5	6	7	8	9	10	11	15	17	19	22
Frequency	13	4	13	6	11	10	8	4	1	5	5	4	2	1	1	1

Illustrate this information as a histogram.

5   The final scores in the 1971 Dunlop Masters Golf Tournament were:

```
273  275  277  277  278  278  278  279
280  280  280  280  281  281  282  282
282  283  284  285  286  290  293  293
293  294  299  300  301
```

Group the scores using a suitable class interval, and illustrate them by means of a histogram.

6   Make a frequency table and histogram for the data given in question *2* of Exercise 2.

7   Illustrate the frequency distribution which you obtained in question *1* of Exercise 3 as a histogram.

5 Frequency polygons

Another way of showing a frequency distribution is by means of a frequency polygon. Here we draw ordinates, or 'uprights', through the midpoints of the class intervals on the horizontal axis, with heights proportional to the frequencies. The lines joining the points at the tops of the ordinates form the frequency polygon.

Figure 7 shows the frequency polygon for the data obtained from the experiment of tossing 9 coins a number of times.

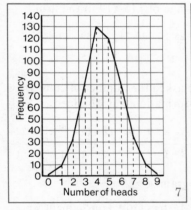
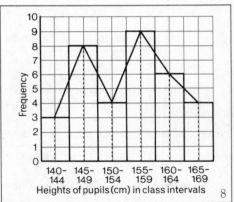

Figure 8 shows the frequency polygon for the data covering the heights of school children, already illustrated by the histogram in Figure 5.

Exercise 5

1. On the same diagram draw a histogram and a frequency polygon for the data below concerning the marks scored in a mathematics test by second-year pupils in a school.

Mark scored	1–10	11–20	21–30	31–40	41–50	51–60	61–70	71–80
Frequency	0	3	10	15	24	28	18	9

Frequency polygons

2 The heights in centimetres of a group of pupils were:

    122 132 145 135 150 157 147 148 154 151
    127 140 148 150 152 150 149 156 152 125
    134 145 147 152 151 147 146 152 150 147
    132 150 145 155 154 150 147 150 149 145

Using an appropriate class interval make a frequency table, and draw a histogram of the heights. Show also the frequency polygon for the distribution.

3 The results of the English League Division I football matches on 16th October 1971 are shown in question *1* of Exercise 2. Illustrate as a histogram the number of goals scored. Show also the frequency polygon for the distribution.

4 A first-year class noted the number of people in cars passing the school one day at lunchtime. Illustrate the information obtained as a histogram and as a frequency polygon.

Number of people in car	1	2	3	4	5	6
Number of cars	35	27	12	22	12	2

What is the relative frequency of 'exactly 1 person in a car' in this sample?

5 The number of blooms on sweet-pea stems is shown in the frequency polygon of Figure 9.

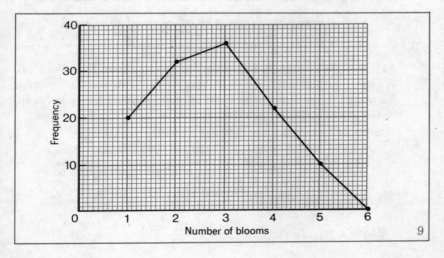

Arithmetic

  a Find the total number of stems.
  b What is the mode of the distribution?
  c What is the relative frequency of a stem with (*1*) exactly 3 blooms (*2*) at least 3 blooms, in the sample?

6 Draw a pie chart and a frequency polygon to illustrate the distribution of 'intelligence' in a section of the population, as indicated by a verbal reasoning test:

Intelligence quotient	55–69	70–84	85–99	100–114	115–129	130–144
Percentage of population	2	14	34	34	14	2

Assuming that this is a typical sample of the population, what is the probability that a person selected at random will have an intelligence quotient of: *a* 130-144   *b* 85-99?

7 *a* Select a book and choose a page at random from it. Count the numbers of letters in each of the first 100 words on the page. Draw up a frequency table of the numbers of letters in the words. Illustrate the information as a histogram and as a frequency polygon.

What is the modal number of letters per word in the passage you have chosen? Compare your results with those of a pupil who chose a different passage.

  b Repeat the question but this time count the number of words in each of the first 100 sentences.

6 Averages–measures of central tendency

Averages are commonly used to compare samples of the same kind. We speak of goal averages, or batting averages, or the class average for marks in an examination, the average wage of the population, or the average rainfall in a certain area.

Suppose two boys had examination marks as follows:

	English	French	History	Mathematics	Science
John	75	80	81	90	84
James	72	84	86	60	56

$$\text{John's average mark} = \frac{75+80+81+90+84}{5} = \frac{410}{5} = 82$$

$$\text{James's average mark} = \frac{72+84+86+60+56}{5} = \frac{358}{5} = 71 \cdot 6$$

Averages—measures of central tendency

John's average mark of 82 compared with James's average mark of 71·6 describes John's greater success in the examinations, without giving the complete list of marks.

The *average* of a collection of numbers or measures found in this way is called the *mean* of the numbers or measures.

$$\text{Mean} = \frac{\text{the sum of all the measures}}{\text{the number of measures}}$$

Other kinds of average used to describe a sample include:
the *mode*, which is the most frequent measure;
the *median*, which is the middle measure in a collection of ordered measures.
For example, the median of the seven numbers 3, 7, 9, 10, 12, 13, 15 is the fourth number, 10. If an eighth number, 17, is included, there is no middle number, and the median is taken to be the mean of the fourth and fifth numbers, i.e. $\frac{1}{2}(10+12) = 11$.

Example. Find the mode, mean and median of the numbers:

$$4, 5, 6, 7, 7, 8, 8, 8, 8, 9, 9, 10, 11, 12, 13$$

Mode = the most frequent number = 8

$$\text{Mean} = \frac{4+5+6+7+7+8+8+8+8+9+9+10+11+12+13}{15}$$

$$= \frac{125}{15} = 8 \cdot 3$$

Median = the middle number = the eighth number in fifteen = 8

The histogram in Figure 10 indicates the reason for these measures of average being called *measures of central tendency* (where the greatest frequency normally occurs).

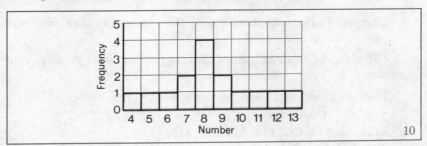

Arithmetic

Exercise 6

1. Calculate the mean of each of the following:
   a. 7, 7, 8, 9, 10, 10, 12
   b. 25 cm, 19 cm, 16 cm, 14 cm, 21 cm
   c. 14 kg, 25 kg, 16·4 kg, 15·1 kg, 19·5 kg
   d. £1·50, £1·05, £1·70, 75p, 34p, 36p.

2. The duration, in minutes, of telephone calls made by a business man on a certain day was:

   3 7 10 2 4 8 11 9 6 3
   2 4 8 15 14 10 8 7 4 7

   Calculate the mean duration of his calls.

3. The number of hours of sunshine recorded daily in a city during the last fortnight of April one year was:

   3·8 7·8 5·7 2·0 3·4 7·2 4·1
   4·9 6·3 0·8 1·3 7·9 7·6 5·2

   What was the average (mean) daily number of hours of sunshine for the city over this period?

4. Find the mean of the following test scores:
   a. 3 4 5 5 4 3 6 7 8 4 5 6 8 9 9 7 5 6 6 8
   b. 6 7 8 5 7 8 9 6 7 8 5 8 9 7 8 6 9 8 5 8

5. Consumers' expenditure on cars and motor-cycles for the four quarters of 1969 and 1970 are listed below in millions of £s.

	1969	1970
First quarter	155	183
Second quarter	182	206
Third quarter	200	219
Fourth quarter	187	237

   Calculate to the nearest million £s the average (mean) expenditure per quarter in: a. 1969 b. 1970.

6. Calculate the mean number of children per family for pupils in your class.

7. Find the mode, mean and median of each of the following:
   a. 2, 2, 3, 3, 3, 4, 4, 4, 4, 5, 5, 6, 7
   b. £12, £13, £13, £13, £14, £15, £15, £16, £17
   c. 2, 9, 1, 2, 5, 7, 2, 3, 1, 4, 4, 8

Averages—measures of central tendency

8 Find the mean score (to the nearest whole number) in the final round of the Dunlop Masters Tournament as detailed in question *3* of Exercise 2.

9 In nine arithmetic tests during a term a boy's scores (out of 25) were: 20, 22, 18, 21, 22, 16, 14, 19, 17.
    Which of the three averages—mode, mean, median—would he prefer to count as his 'mark'?

Exercise 6B

1 A batsman scored a total of 588 runs in 24 matches. Find his average score per match.

2 The number of hours of sunshine at a seaside resort in seven successive days were:
    7·3, 4·8, 1·7, 6·4, 5·9, 7·6, 6·9.
Calculate the mean daily number of hours of sunshine during this period.

3 The members of a school's relay team weighed 47·5 kg, 49 kg, 52 kg and 53·5 kg. Find their mean mass.

4 The number of new car registrations (in thousands) are given below.

	1969	1970
First quarter	233	250
Second quarter	253	275
Third quarter	267	291
Fourth quarter	235	281

Calculate to the nearest thousand the average (mean) number of registrations per quarter in: *a* 1969 *b* 1970.

5 The average weekly wage in a factory employing 125 people was £32. What was the total amount paid in wages weekly?

6 The mean height of 4 boys was 152 cm. When a fifth boy joins the group the mean height is increased by 2 cm. What is the height of the fifth boy?

7 In a housing development of 600 houses, the houses have a mean frontage to the street of 15·2 metres, and they occupy a mean ground area of 425 square metres. Find the total length of frontage and the total ground area occupied by the houses.

Arithmetic

8. During one season a football team played 27 games, and scored the following number of goals per match. The frequency is given in brackets:

    0 (5), 1 (7), 2 (4), 3 (6), 4 (3), 6 (2)

    Calculate the mean, median and modal score.

9. During the course of one year the distance travelled by a car was 13 563 km, and a total of 1480 litres of petrol was used. Find the average performance of the car in kilometres per litre, rounded off to 1 decimal place.

10. In an International Rugby Match the average mass of one set of forwards was 92·5 kg and of the other set of forwards 87·25 kg. By how much did the total mass of the one set exceed that of the other set? (There are 8 forwards on each side.)

11. The average monthly deficit in a country's trade balance for the first 11 months of a particular year was £68 million. In December it was £107 million. What was the average monthly deficit over the whole year, to the nearest million pounds?

7 Mean from a frequency table

We can extend the frequency table to help us to calculate the mean as shown in the following example.

The number of goals scored by English First Division football teams on 16th October 1971 was:

1, 2, 1, 1, 3, 0, 2, 0, 1, 0, 1, 2, 2, 3, 3, 2, 4, 1, 0, 0, 1, 0

To calculate the mean number of goals scored per team we construct the following table:

Number of goals		Frequency	Number of goals × frequency
0	ⅢⅢ Ⅰ	6	0
1	ⅢⅢ ⅠⅠ	7	7
2	ⅢⅢ	5	10
3	ⅠⅠⅠ	3	9
4	Ⅰ	1	4
	Totals	22	30

Mean number of goals = $\frac{30}{22}$ = 1·4, to 2 significant figures.

Mean from a frequency table

Exercise 7

1. The football scores for the Scottish League Division I on 16th October 1971 were:

Aberdeen	2	Hibernian	1	Falkirk	2	Dunfermline	1
Ayr Utd	0	Clyde	1	Hearts	1	Airdrie	1
Celtic	3	Dundee	1	Motherwell	3	Morton	1
Dundee Utd	1	Rangers	5	Partick	2	St. Johnstone	1
East Fife	2	Kilmarnock	0				

   Make a frequency table of the goals scored by the teams, and hence calculate the mean score to 2 significant figures.

2. Make a frequency table of the following marks scored in a test (out of 25), and hence calculate the mean mark.

   ```
   10  18  17  12  24  16  16  14  20  18
   13  16  22  13  17  17  17  21  14  15
   19  16  14  17  21  15  16  19  18  16
   ```

3. Two classes were asked how many textbooks they had with them; the results are shown below. Make a frequency table, and calculate the mean number of books per pupil.

Number of textbooks	0	1	2	3	4	5	6	7	8	9	10
Frequency	0	1	1	7	9	10	10	11	8	2	2

4. The following table gives the number of textbooks carried by each pupil in a certain class on a certain day.

   ```
   4  12   2   8   9   9   4   2  10   3
   8   3   9   6  11   3  10   2   6  11   5
   8   6   2   5   8  10   1   3   8   9  11   6
   ```

   Draw up a frequency table showing the number of pupils carrying 1, 2, 3, ... books and calculate the mean number of books carried per pupil. Draw a histogram to illustrate the distribution. Can you suggest an explanation for the unusual shape of this distribution? Calculate to the nearest whole number the percentage of pupils:

   a. who were carrying fewer than 4 books
   b. who probably did not go home for lunch that day.

5. A machine is designed to pack a commodity in 50-gramme cartons. As a check on its accuracy a sample of packets is taken at regular intervals, and the packets are weighed to the nearest gramme. The histogram in Figure 11 shows the distribution obtained.

Arithmetic

a How many packets are there in the sample?
b Make a frequency table, and calculate the mean mass of the packets (to the nearest gramme).
c What is the relative frequency of a packet in the sample having a mass of 52 g?
d Assuming that this sample is typical of the population of all such packets, what is the probability that a packet chosen at random will have a mass of less than 50 g?

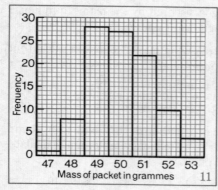
11

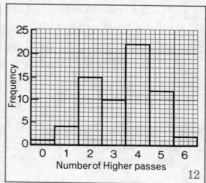
12

6 The histogram shown in Figure 12 illustrates the distribution of Higher Grade passes obtained by fifth-year students in a school last year. Use the histogram to answer the following questions:

a How many students were there in the fifth year?
b Make a frequency table, and calculate the mean number of Higher passes obtained by the students (rounded off to 1 decimal place).
c What was the modal number of Higher passes obtained?
d What was the relative frequency of 5 or more Higher passes?

7 Calculate the mean number of coins the pupils in your class have with them.

8 Calculate the mean height, to the nearest cm, of pupils in your class. Illustrate the distribution by means of a histogram.

Summary

1 *Population, Sample*

The set of items which have the characteristic in which we are interested is called a *population* and the subset of these items actually measured is called a *sample* of the population.

2 *Frequency table*

Measure		Frequency
3	///	3
4		0
5	++++ //	7
6	++++	5
	Total	15

3 *Histogram, Frequency Polygon*

In a histogram the area of each rectangle is proportional to the frequency.

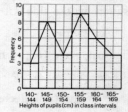

4 *Averages*

$$\text{Mean} = \frac{\text{the sum of all the measures}}{\text{the number of measures}}$$

Mode = the most frequent measure

Median = the middle measure in a collection of ordered measures.

5 *Mean from a frequency table*

Number of goals		Frequency	Number of goals x frequency
0	++++ /	6	0
1	++++ //	7	7
2	++++	5	10
3	//	2	6
	Totals	20	23

Mean number of goals = $\frac{23}{20}$ = 1·15

Revision Exercises

Revision Exercise on Chapter 1
Squares and Square Roots of Numbers

1. Write down the squares of the following numbers:
   a 8   b 12   c 0·7   d 1·4   e 0·05

2. Which of the following numbers are exact squares?
   a 81   b 2·5   c 360   d 121   e 1·21

3. Write down all the whole numbers between 20 and 90 which are squares of whole numbers.

4. Find the area of a square whose side is:
   a 3 cm   b 4 mm   c 60 m   d 5·8 m   e 25 km

5. Use the graph of Figure 1, page 162, to find approximations for:
   a $2·4^2$ b $6·7^2$ c $8·3^2$ d $1·75^2$ e $5·55^2$

6. Use tables or slide rule to find (to 2 decimal places):
   a $2·3^2$ b $6·9^2$ c $8·73^2$ d $5·86^2$ e $9·09^2$

7. Write the following numbers in the form $a \times 10^n$ and then, using the tables of squares, find their squares.
   a 28·4   b 456   c 0·34   d 0·246   e 0·087

*   *   *

8. Write down the square roots of:
   a 9   b 144   c 1600   d 225   e 6·25

9. Find the side of a square whose area is:
   a 36 m²   b 2500 cm²   c 1·44 mm²   d 289 m²   e 1·96 cm²

10. Find the perimeter of a square whose area is 1225 cm².

11. Which of the following numbers have exact square roots?
    a 4   b 90   c 0·4   d 0·01   e 1·6

Revision Exercise on Chapter 1

12 Between which two consecutive whole numbers do the following lie? (e.g. $3 < \sqrt{10} < 4$).

   a $\sqrt{91}$ *b* $\sqrt{27}$ *c* $\sqrt{4 \cdot 9}$ *d* $\sqrt{411}$

13 Estimate to 2 significant figures:

   a $\sqrt{13}$ *b* $\sqrt{3}$ *c* $\sqrt{58}$ *d* $\sqrt{70}$ *e* $\sqrt{31}$

14 Using tables and/or slide rule find the square roots of:

   a 171   *b* 2900   *c* 0·92   *d* 5·79   *e* 0·067
   f 2389   *g* 0·0022   *h* 1·08   *i* 1080   *j* 10800

15 First obtain a good estimate and then use the iterative method to find the square roots of the following numbers, rounded off to 3 significant figures.

   a 11   *b* 18   *c* 89   *d* 6·3   *e* 2·47

16 By using tables or slide rule first, and then the method of iteration, calculate to 5 significant figures:

   a $\sqrt{27}$ *b* $\sqrt{73}$ *c* $\sqrt{5 \cdot 8}$ *d* $\sqrt{1 \cdot 62}$ *e* $\sqrt{96}$

17 A rectangular field measures 150 m by 54 m. Find the length of the side of a square which has the same area.

18 The total area of all the faces of a cube is 294 cm². Find the volume of the cube.

19 Write down the 5th, 8th and 20th terms of a sequence of whole numbers which starts 1, 4, 9, 16, ...
Which term of the sequence is one million?

20 Given that $\sqrt{7} \doteqdot 2 \cdot 65$ and that $\sqrt{70} \doteqdot 8 \cdot 37$ find the approximate square roots of:

   a 700   *b* 7000   *c* 700000   *d* 0·07   *e* 0·7

21 Given that $\sqrt{40} = 6 \cdot 32$ and knowing $\sqrt{4}$ find the following square roots, saying which are approximate and which are exact:

   a $\sqrt{4000}$ *b* $\sqrt{0 \cdot 4}$ *c* $\sqrt{400}$ *d* $\sqrt{0 \cdot 04}$ *e* $\sqrt{400000}$

22 If a rectangle is four times as long as it is broad, show in a sketch how it can be divided into four equal squares. If the area of the rectangle is 784 cm², find the lengths of its sides.

   If a square is equal in area to the rectangle, write down the length of its side.

Arithmetic

23 A rectangle, twice as long as it is broad, has an area of 10 cm². Find its dimensions as accurately as your tables allow.

24 Find the values of p^2 and $\sqrt{p}$ when p is replaced by:
   a 7·3   b 73   c 0·6   d 600

25 A cuboid has a square base of side x cm, height 8 cm, and volume 18 cm³. Find x.

26 If $r = \sqrt{\dfrac{7A}{22}}$ find r when A has the value
   a 154   b 11   c 20

27 If $q = x^2\sqrt{x}$ calculate q when x has the value
   a 9   b 100   c 10

Revision Exercise on Chapter 2
Using a Slide Rule

1 Find rough estimates for:

a	2·34 × 3·71	b	18·7 ÷ 6·29	c	8·91 × 20·9
d	48·9 ÷ 374	e	0·216 × 0·873	f	0·896 ÷ 0·475
g	823 × 0·164	h	823 ÷ 0·164	i	82·3 × 16·4
j	8·23 ÷ 16·4	k	0·764 × 0·435 × 0·987		

Use a slide rule to calculate the following:

2 8·7 × 13·4   3 8·7 ÷ 13·4   4 6·28 × 0·896

5 0·896 ÷ 6·28   6 586 × 427   7 0·586 ÷ 0·427

8 Express as percentages to the nearest whole number:

   a $\dfrac{8}{13}$ b $\dfrac{£3·76}{£16·50}$ c $\dfrac{48\text{p}}{£2·57}$ d $\dfrac{35 \text{ cm}}{2·36 \text{ m}}$

9 $0·723 \times 0·456 \times 3·14$ 10 $\dfrac{573 \times 285}{273}$

11 $\dfrac{28·3 \times 0·125}{576}$ 12 $\dfrac{1·57}{47·5 \times 8·9}$

Revision Exercise on Chapter 3

13 $\left(\dfrac{39.7}{58.6}\right)^2$

14 $\sqrt{\dfrac{8.63}{3.14}}$

15 $1.09 \times \dfrac{295}{273} \times \dfrac{760}{748}$

16 $\dfrac{0.916 \times 0.213}{0.990 \times 0.568}$

Revision Exercise on Chapter 3
The Circumference and Area of a Circle

1 Calculate the circumference and area of a circle of radius 70 m.

2 The diameter of the circular base of a lampshade is 50 cm. What is its circumference? How much would it cost for braid round the base at 60p per metre?

3 The moon is approximately 385 000 km from the centre of the earth. Assuming that it follows a circular path, how far does the moon travel in going once round the earth?

4 A circular flower-bed has a diameter of 14 metres. What is its area? How much fertilizer, to the nearest necessary kilogramme, must be bought in order to apply it at the rate of 50 grammes per square metre?

5 Figure 1 shows a rectangle from which four quarter circles have been removed. Calculate the perimeter and area of the remaining shape.

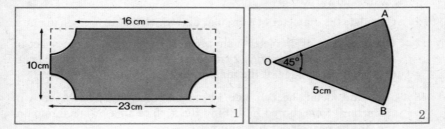

6 Figure 2 shows an arc AB of a circle, centre O, of radius 5 cm. Angle AOB is 45°. Calculate the perimeter and area of the shape OAB.

Arithmetic

7   The diameter of the top of a stool is 40 cm. Calculate its surface area.

8   A rectangular lawn 12 m long and 7 m broad has two circular flower-beds cut from it, each of diameter 2 m. Find the area of grass to the nearest square metre.

9   A wheel of a motor-car has a diameter of 35 cm. Find the circumference of the wheel, and the distance travelled by the car as the wheel makes 100 turns.

10 A merry-go-round consists of two rings of horses which make four complete turns per minute. The radius of the outer ring is 14 m, and of the inner ring is 10·5 m. Calculate the distance the horses in each ring travel during one circuit, and hence calculate the speed of a horse in each ring.

11 A circular pool of diameter 26 m has a path 1 m wide round its edge. By subtracting two circular areas find the area of the path to the nearest square metre.

12 To make lids for tin cans the largest possible circular discs are cut from a rectangular strip of tin 240 cm long and 7 cm broad. If as many discs as possible are cut out, how much material is wasted?

13 Calculate the radius of a circle, given that:
  a the circumference is 143 cm   *b* the area is 157 m².

14 Find the diameter of a pond if its circumference is 176 m.

15 A circular jigsaw puzzle has a circumference of 88 cm. Calculate its radius and surface area.

16 Calculate the diameter of a circular disc which has an area of 86 mm².

17 Circular discs of diameter 2 cm are cut from a rectangular sheet of cardboard 90 cm long and 12 cm broad. How many discs can be cut; and what percentage of the material is then wasted?

18 A square metal sheet of side 10 cm has 16 circular holes, each of diameter 0·5 cm, bored in it. Find the surface area of metal remaining, to the nearest square centimetre.

19 A satellite is moving in a circular orbit 2620 km above the earth's surface, and completes each orbit in 2 hours. Assuming that the radius of the earth is 6380 km, find the speed of the satellite in km/h.

Revision Exercise on Chapter 4

20 A satellite has a speed of 16 500 km/h, and makes one complete orbit of the earth in 6 hours. Assuming that the orbit is circular, and that the radius of the earth is 6400 km, calculate the height of the orbit above the earth.

Revision Exercise on Chapter 4
Statistics — 2

1 The numbers of eggs collected daily over a period of six weeks by a smallholder were:

```
76  80  93  84  88  97   83  100  94   77  97
80  88  90  79  85  99  100   96  91   82  88
93  98  99  89  81  86   95   94 102   94
88  92 101  80  90  81   91   95  87   93
```

  a Construct a frequency table with a class interval of 5, starting at 76.
  b What would the frequencies be if a class interval of 10 were chosen?

2 From a survey made on a new housing estate, the distribution of the ages of the children under 5 years is shown in the accompanying histogram (Figure 3).

  a How many children under 5 years of age are there on the estate?

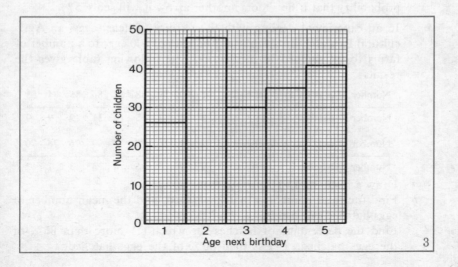

3

Arithmetic

    b Calculate, to the nearest whole number, the percentage of the children born in the twelve months preceding the survey.

    c Represent the information given in the histogram on a pie chart, of radius 3 cm, labelling each compartment with the number denoting the age next birthday of the children in the group.

3 A business has 50 employees who work a five-day week. The manager compiled the following data of the number of days lost through illness in one period of six months:

Number of days lost	1	2	3	4	5	9	22
Number of employees who lost this number of days	8	5	7	0	6	1	1

    a How many employees had no absence due to illness over the period?
    b What percentage of employees were absent for less than 4 days?

4 In an archery tournament each competitor in a York Round shoots 72 arrows at the target 100 yards distant. Gold scores 9 points, red 7, blue 5, black 3, white 1, and a miss 0. A competitor's scores are:

```
0 3 1 0 5 3 3 0 1 3 3 1 5 0 0 3 1 3
5 3 0 7 3 3 1 3 5 9 5 7 0 3 1 1 3 3
1 0 3 5 1 3 1 5 3 0 3 1 3 3 5 3 7 3
0 1 3 3 0 5 3 9 5 3 0 1 5 3 1 7 3 3
```

    a Make a frequency table and calculate his mean score.
    b Assuming this to be a typical sample of his shooting, what is the probability that if he shoots another arrow it will score 5?

5 In an experiment dealing with the fertility of hens' eggs, an Agricultural Research Institute issued batches of 100 eggs to a number of farms to be hatched in incubators. The following table gives the results:

Number of eggs hatched per batch	91	90	89	88	87	86	85	84	83
Number of batches	3	1	1	5	12	11	9	9	9

Number of eggs hatched per batch	82	81	80	79	78	77	76	75	70
Number of batches	9	9	3	5	4	4	3	1	2

    a Draw a histogram to illustrate these results.
    b Find the total number of batches issued, and the mean number of eggs hatched per batch.
    c Find the percentage of batches for which (*1*) more than 86% of the eggs hatched (*2*) less than 80% of the eggs hatched.

Revision Exercise on Chapter 4

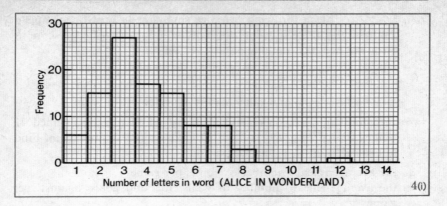

4(i)

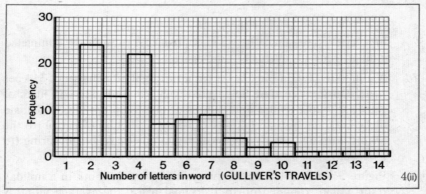

4(ii)

6   Each of the two books *Alice in Wonderland* and *Gulliver's Travels* was opened at random, and the numbers of letters in each of the first hundred words on the page were counted. The results of this survey are shown in Figure 4: (i) refers to *Alice* and (ii) to *Gulliver*.

   Calculate, for the chosen sample in each book:
   a   The modal number of letters per word.
   b   The mean number of letters per word.
   c   The percentage of words which contain 2, 3, or 4 letters.
   d   The percentage of words which have more than 6 letters.
        Which of these books would you judge (from these statistics) to be the easier to read?

7   Write out as an array the set of all possible outcomes on throwing two dice simultaneously. Hence calculate the probabilities of scoring

Arithmetic

2, 3, 4, ..., 12, in throwing two dice simultaneously. Now complete the following table showing the expected frequency of each score in throwing two dice simultaneously 144 times:

Score	2	3	4	5	6	7	8	9	10	11	12
Frequency											

Draw a histogram and calculate the expected mean score.

8 In a football season a certain team played 34 matches, gained 41 points, scored 63 goals and had 37 goals scored against them. Find (all rounded off to 1 decimal place):

a the average (mean) number of points per game
b the average (mean) number of goals 'for' and goals 'against' per game
c the 'goal average', i.e. $\frac{\text{goals for}}{\text{goals against}}$

9 In a cricket season a batsman scored 859 runs in 31 completed innings. Find his average to the nearest run.

10 A visitor to Britain hired a car for a fortnight at a charge of £14 per week. During that time he covered 5520 km and used 485 litres of petrol at 7·2p per litre and 2 litres of oil at 26p per litre. Calculate the average number of kilometres covered per litre of petrol (to the nearest km) and the average cost per kilometre of his motoring (to 1 decimal place).

11 Figure 5 shows a bar chart giving rainfall each month in a holiday resort. Find, rounded off to 1 decimal place, the average monthly rainfall in centimetres:

a for the whole year
b for the holiday months of June to September inclusive.

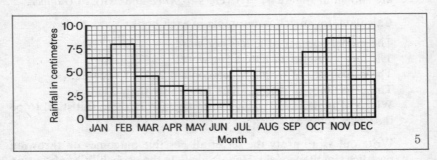

Cumulative Revision Section (Books 1-4)

Book 1 Chapter Summaries

Chapter 1 The system of whole numbers

The set of *whole numbers* is $\{0, 1, 2, 3, 4, ...\}$
The set of *natural numbers* is $\{1, 2, 3, 4, ...\}$
The set of *even numbers* is $\{0, 2, 4, 6, ...\}$
The set of *odd numbers* is $\{1, 3, 5, 7, ...\}$
The set of *prime numbers* is $\{2, 3, 5, 7, 11, 13, 17, 19, ...\}$
   Prime numbers are divisible only by themselves and by 1.

1 *The Commutative Laws of addition and multiplication*
$$a+b = b+a \qquad a \times b = b \times a$$
e.g. $8+24 = 24+8$ \qquad e.g. $8 \times 24 = 24 \times 8$

2 *The Associative Laws of addition and multiplication*
$$(a+b)+c = a+(b+c) \qquad (a \times b) \times c = a \times (b \times c)$$
e.g. $(6+4)+3 = 6+(4+3)$ \qquad e.g. $(6 \times 4) \times 3 = 6 \times (4 \times 3)$

3 *The Distributive Law*
$$(a \times b)+(a \times c) = a \times (b+c) \quad \text{or} \quad a(b+c) = ab+ac$$
e.g. $(7 \times 3)+(7 \times 5) = 7 \times (3+5)$

4 *The identity element for addition* (0)
$$a+0 = 0+a = a$$
e.g. $5+0 = 0+5 = 5$

5 *The identity element for multiplication* (1)
$$a \times 1 = 1 \times a = a$$
e.g. $8 \times 1 = 1 \times 8 = 8$

6 *Squares and cubes*
a^2 means $a \times a$ \qquad a^3 means $a \times a \times a$
e.g. $9^2 = 9 \times 9 = 81$ \qquad e.g. $5^3 = 5 \times 5 \times 5 = 125$

7 *The principle for multiplying by* 0
$a \times 0 = 0 = 0 \times a$ \qquad If a, b are whole numbers for which
e.g. $7 \times 0 = 0 = 0 \times 7$ \qquad $a \times b = 0$, then at least one of a, b is 0.

Arithmetic

Chapter 2 Decimal systems of money, length, area, volume and mass

1 *Length*
1 cm = 10 mm
1 m = 100 cm = 1000 mm
1 km = 1000 m

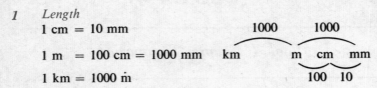

2 *Area*
1 cm² = 100 mm²
1 m² = 10000 cm²   For a rectangle, $A = lb$
1 km² = 1 000 000 m²   For a square, $A = l^2$

3 *Volume*
1 litre = 1000 ml = 1000 cm³

1 m³ = 1 000 000 cm³ = 1000 litres
For a cuboid, $V = lbh$, or $V = Ah$

4 *Mass*
1 g = 1000 mg
1 kg = 1000 g

Chapter 3 Fractions, ratios and percentages

The meaning of a fraction

1 $\dfrac{a}{b} = a \div b$ *a* is the **numerator** of the fraction
     b is the **denominator** of the fraction

2 *Equal fractions*
$$\frac{1}{2} = \frac{1 \times 3}{2 \times 3} = \frac{3}{6} = \frac{3 \times 4}{6 \times 4} = \frac{12}{24} = \frac{12 \div 2}{24 \div 2} = \frac{6}{12} = \ldots$$

3 *Least common multiples*
The LCM of   *a* 2, 3, is 6   *b* 2, 3, 4 is 12   *c* 5, 10 is 10

4 *Addition and subtraction of fractions*
$\frac{3}{4} + \frac{5}{6} = \frac{9}{12} + \frac{10}{12} = \frac{19}{12} = 1\frac{7}{12}$

5 Multiplication of fractions

$$\frac{3}{4} \text{ of } \frac{5}{6} = \frac{\overset{1}{\cancel{3}}}{4} \times \frac{5}{\underset{2}{\cancel{6}}} = \frac{5}{8}$$

6 Division of fractions

$$3\tfrac{1}{3} \div 2\tfrac{1}{2} = \frac{10}{3} \div \frac{5}{2} = \frac{\overset{2}{\cancel{10}}}{3} \times \frac{2}{\underset{1}{\cancel{5}}} = \tfrac{4}{3} = 1\tfrac{1}{3} \qquad \frac{p}{q} \div \frac{a}{b} = \frac{p}{q} \times \frac{b}{a}$$

7 Ratios

$$a:b = \frac{a}{b}$$

8 Percentages

$$1 \text{ per cent} = 1\% = \frac{1}{100} \qquad a\% = \frac{a}{100} \qquad \frac{x}{y} = \frac{x}{y} \times 100\%$$

Book 2 Chapter Summaries

Chapter 1 Decimals and the Metric System

1 Notation

H	T	U	t	h			
		4	6		=	4·6	= $4\tfrac{6}{10}$
1	0	5	7	9	=	105·79	= $105\tfrac{79}{100}$
				3	=	0·03	= $\tfrac{3}{100}$

2 Measurement

All measurements are approximate.

A length of 3·3 cm means that the length is 3·3 cm to the nearest tenth of a centimetre.

3 Approximation

Rounding off

If the next figure is greater than 5, increase the round-off figure by 1.

If the next figure is 5, round off to the nearest even number; e.g. 36·75 = 36·8 to 1 *decimal place*, or to 3 *significant figures*.

Arithmetic

4 *Addition and subtraction*
Examples:

```
  53·40         53·40
+ 31·82       − 31·82
───────       ───────
  85·22         21·58
```

5 *Multiplication*

a To multiply by 10, move the figures one place left (*or* move the decimal point one place right), making the number 10 times greater. Similarly, two places for 100; and so on.

b Example. $47·6 \times 4·3 = 204·68$

```
                        47·6  (1 figure after point)
                         4·3  (1 figure after point)
Estimate:  50 × 4 = 200  ────
                        1428
                       19040
                       ──────
                      204·68  (2 figures after point)
```

6 *Division*

a To divide by 10, move the figures one place right (*or* move the decimal point one place left), making the number 10 times smaller. Similarly, two places for 100; and so on.

b Example. $0·276 \div 4·3$ Estimate

$$= \frac{0·276}{4·3} \qquad \frac{0·3}{4}$$

$$= \frac{2·76}{43} \qquad \doteqdot 0·07$$

$$= 0·064\ldots$$

```
        0·064
    43)2·76
       258
       ───
       180
       172
       ───
         8
```

7 *Expressing a common fraction as a decimal*

$\frac{5}{7} = 0·714$, to 3 significant figures.

```
   0·7142
7)5·0000
```

8 *Standard form*

$a \times 10^n$ where $1 \leqslant a < 10$, and $n \in N$.

$2300 = 2·3 \times 10^3$

$608\,000 = 6·08 \times 10^5$

$12\,478 = 1·25 \times 10^4$, to 3 significant figures.

9 *The metric system*

km —(1000)— m —(1000)— cm —(100)— mm, cm —(10)— mm

Length: 1 cm = 10 mm
1 m = 100 cm = 1000 mm
1 km = 1000 m
Volume: 1 litre = 1000 ml = 1000 cm³
Mass: 1 g = 1000 mg
1 kg = 1000 g
1 tonne (metric ton) = 1000 kg

10 Calculating a percentage of money

Example. $4\frac{1}{2}\%$ of £6·21

$= \frac{4\frac{1}{2}}{100} \times £6\cdot 21$

$= 4\frac{1}{2} \times £0\cdot 0621$

= 28 p, to the nearest penny

$$\begin{array}{r} £0\cdot 0621 \\ \times 4 \\ \hline 0\cdot 2484 \\ 0\cdot 0310 \\ \hline 0\cdot 2794 \end{array}$$

Chapter 2 Computers and Binary Arithmetic

1 *a* The binary system has only two figures, 0 and 1, and is a *base two* system.

b Notation   S E F T U   T U
    $1\,0\,1\,0\,1_{two} = 16+4+1 = 2\,1_{ten}$

c Addition

+	0	1
0	0	1
1	1	10

d Multiplication

×	0	1
0	0	0
1	0	1

2 *a* Computers can:

(i) store information
(ii) carry out calculations very rapidly

b Punched cards and tape can be used to programme a computer by means of binary patterns of holes.

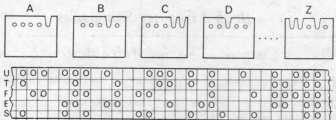

Arithmetic

Chapter 3 Statistics—1

Statistical data are shown in this chapter by means of:

1. **Pictographs, e.g.**

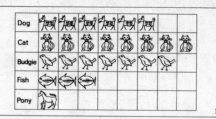

2. **Bar charts, e.g.**

3. **Pie charts, e.g.**

4. **Line graphs, e.g.**

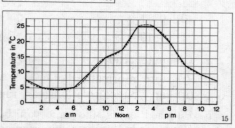

Book 3 Chapter Summaries

Chapter 1 Social Arithmetic—1

1. *Electricity and gas accounts* can be checked by calculation.
2. *Ready reckoners* are used to save time when certain calculations are repeated.

3 A *discount* of 20% on £50 = $\frac{1}{5} \times £50$ = £10;
    of $3\frac{1}{2}$% on £50 = £1·75.

$$\begin{array}{rcl} 1\% \text{ of } £50 &=& £0·50 \\ 3\% \text{ ,, ,, } &=& 1·50 \\ \tfrac{1}{2}\% \text{ ,, ,, } &=& 0·25 \\ \hline 3\tfrac{1}{2}\% \text{ ,, ,, } &=& £1·75 \end{array}$$

4 *Banking.* Cheques can be used with current accounts.
    Interest is paid on deposit (savings) accounts.
    A *rate of interest* of 5% per annum gives interest of £5 in a year on a principal of £100.

5 *Profit and loss.* Percentage profit or loss may be calculated on the cost price or on the selling price.

6 *Percentages.* Some useful common fractions, decimal fractions and percentages:

$10\% = \tfrac{1}{10} = 0·1$ $33\tfrac{1}{3}\% = \tfrac{1}{3} = 0·33$ (to 2 sig. figs.)
$20\% = \tfrac{1}{5} = 0·2$
$25\% = \tfrac{1}{4} = 0·25$ $66\tfrac{2}{3}\% = \tfrac{2}{3} = 0·67$ (to 2 sig. figs.)
$50\% = \tfrac{1}{2} = 0·5$
$75\% = \tfrac{3}{4} = 0·75$ $100\% = 1$

Chapter 2 Ratio and Proportion

First variable   *Second variable*
  $a \longleftrightarrow c$
  $b \longleftrightarrow d$

1 If the ratios $\dfrac{a}{b}$ and $\dfrac{c}{d}$ are equal, or

if the ratios $\dfrac{a}{c}$ and $\dfrac{b}{d}$ are equal, then the variables are related by *direct proportion*.

Corresponding variables increase or decrease in the *same ratio* (e.g. if one is doubled, the other is doubled).

The graph showing the relationship between the variables consists of a set of points which may be joined by a straight line passing through the origin.

Arithmetic

2. If $\dfrac{a}{b} = \dfrac{d}{c}$, i.e. $ac = bd$, the variables are related by *inverse proportion*.

   As one variable increases, the corresponding one decreases in the *inverse ratio* (e.g. if one is doubled, the other is halved).

3. The *scale of a map* is often given by the *representative fraction*, i.e.
$$\frac{\text{distance on the map}}{\text{corresponding distance on the ground}}.$$

Chapter 3 Introduction to Probability

1. The *probability* of an outcome in a random experiment is the limit of the relative frequencies of the outcome in large numbers of trials.

   Where an experiment has several *equally likely* outcomes, the probability of a 'favourable outcome'
$$= \frac{\text{the number of 'favourable' outcomes}}{\text{the number of possible outcomes}}$$

2. If the probability of an outcome of an experiment is P, then the probability that the outcome will not happen is $1-\text{P}$.

3. $0 \leqslant \text{P} \leqslant 1$.

4. In a number of trials the expected frequency of an outcome = the probability of the outcome × the number of trials.

5. For two mutually exclusive outcomes A and B of an event,
$$\text{P}(A \text{ or } B) = \text{P}(A) + \text{P}(B)$$

6. For two independent outcomes A and B of an event,
$$\text{P}(A \text{ and } B) = \text{P}(A) \times \text{P}(B)$$

Summary

Chapter 4 Times, Distance, Speed

1. A *time* given as 09 53 means
   9.53 am, or 7 minutes to 10 in the morning.

   A *time* given as 16 30 means
   4.30 pm, or half past 4 in the afternoon.

2. $\text{Average Speed} = \dfrac{\text{Distance}}{\text{Time}}$

3. If D represents distance, S represents speed and T represents time,
   $$D = ST$$

Cumulative Revision Exercises

Exercise A

1. a Calculate the cost of 180 books at 34p each.
   b Divide £7·48 by 12 to the nearest penny.

2. Express as a product of prime factors: *a* 42 *b* 24 *c* 120.

3. Write down: *a* the set of multiples of 8 which are less than 100
   b the set of multiples of 14 which are less than 100. Hence find the LCM of 8 and 14.

4. The symbol $\odot$ denotes the operation 'double the first number and then add the second number'. Evaluate:

   a $7 \odot 5$ and $5 \odot 7$ *b* $(3 \odot 5) \odot 2$ and $3 \odot (5 \odot 2)$.

   The symbol $*$ denotes the operation 'double the first number and then multiply by the second number'. Evaluate:

   c $7 * 5$ and $5 * 7$ *d* $(3 * 5) * 2$ and $3 * (5 * 2)$.

   Does it seem that the commutative and associative properties are true for the operations $\odot$ and $*$?

5. Arrange in order of size, smallest first:

   a $\tfrac{5}{12}, \tfrac{3}{8}, \tfrac{1}{3}$ *b* $\tfrac{7}{31}, \tfrac{7}{33}, \tfrac{7}{32}$

6. Multiply $1\tfrac{3}{4}$ by:

   a 4   *b* 20   *c* 60   *d* $\tfrac{1}{8}$ *e* $1\tfrac{3}{4}$

Arithmetic

7 Evaluate:

a $(0.2)^2 \times 0.09$

b $9.65 - 12.83 + 7.76$

8 Express in standard form:

a 327 b 45000 c 8 million d 923.6

9 In Figure 1, four towns are represented by A, B, C, and D, and the distances in kilometres between them are shown.

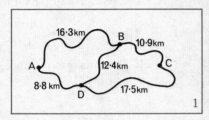

a How many possible routes are there from A to C, no town being included more than once in any route?

b What is the difference in distance between the longest and the shortest of these routes from A to C?

10 Find the average of £6·85, £7·93, £8·84, £5·69 and £9·04.

11 A school received a consignment of twelve calculating machines, each packed in a box weighing 1·1 kg. The total mass of the machines and boxes was 69·6 kg. Calculate the mass of each machine.

12 A rectangular sheet of plastic is 5·5 m long and 1·5 m broad. What will it cost at 56p per square metre?

What would be the cost of a sheet three times as long and half as broad?

13 A rectangular hall 6 m long and 1·6 m broad is to be completely covered with square tiles, each of side 15 cm. How many tiles must be bought? How many of the tiles will have to be cut? If the tiles cost £1·80 per box of 36, or 5½p each if bought separately, find the total cost.

14 Figure 2 shows the shape of the floor of a room. Calculate the cost of covering it with carpet at £3·20 per m².

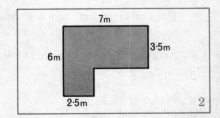

Cumulative Revision Exercises   241

15 Calculate, to three significant figures, the volume of a cuboid 16·8 cm long, 12·5 cm broad and 7·8 cm high.

16 A brick wall is to be 6 m long, 1·2 m high and 15 cm thick. How many bricks, each 25 cm by 12 cm by 7·5 cm, will be required for the wall? (Ignore the thickness of cement.)

17 Which of the following are approximations for $\sqrt{745}$?
   a 86·3   *b* 27·3   *c* 8·63   *d* 2·73 × 10

18 Calculate the area of a rectangle which measures 6·5 cm by 2·6 cm. Calculate also the perimeter of a square equal in area to the rectangle.

19 Find the rate for each of the following in the units given:
   a I travelled 3600 km and used 450 litres of petrol (km per litre).
   b An electric fire used 84 units of electricity in 168 hours (units per hour).
   c A hose delivered 4536 litres of water in 42 minutes (litres per minute).

20 When petrol cost 6p per litre the cost of filling a tank was £1·26. The cost of petrol has risen by 1p per litre. What is the cost of filling the tank now?

21 When three boys stand on a weighing machine platform the machine registers 126 kg. What would it register if five boys stood on the platform?

22 14 Swedish kronor are worth 9·2 German marks. How many marks (rounded off to the second decimal place) are equivalent to 100 kronor?

23 A map has a scale of 1:200000. A rectangular park measures 0·35 cm by 0·15 cm on the map. What are the dimensions of the park in metres? Find the area of the park in hectares given that 1 hectare = 10000 m².

24 A transparency 2·5 cm long and 1·9 cm broad is projected on a screen. If the picture is 2 m long, what is its breadth?

25 12 Argentinian pesos are approximately equal to 100 British pence. Draw a graph for converting pesos to pence (up to 100 pence). Use it to convert:
   a 5 pesos, 7 pesos, 2·5 pesos to pence
   b 80 pence, 27½ pence, 66 pence to pesos

Arithmetic

26 Express all the prime numbers between 30_{ten} and 40_{ten} in binary notation.

27 Calculate $1+11$, $1+11+111$, and $1+11+111+1111$, all the numbers being in binary form.

28 Write down the next two terms of each of the sequences starting with the following numbers in the binary scale. Then express all the terms in the denary (decimal) scale:

   a 1, 10, 100, 1000, ...   *b* 1, 10, 101, 1010, ...

29 How many pieces of wire each 1101_{two} cm long can be cut from a length of 110110_{two} cm of wire? Check your working by converting all the numbers to decimal form.

30 Find the total cost of the following:

$12\frac{1}{2}$ m of stair carpet at £2·20 per m
$4\frac{3}{4}$ litres of stain at 64p per litre
28 carpet clips at 75p per dozen.

31 Calculate the total cost of:

6000 sheets of loose-leaf paper at 40 sheets for 7p
1500 exercise books at 25p per dozen
200 textbooks at $47\frac{1}{2}$p each.
Allow 4% discount on the total.

32 Two successive readings of an electricity meter were 4783 and 5163. The first 36 units are charged at 2·5p per unit, the next 64 units at 1·25p per unit, and the rest at 0·55p per unit. Calculate the cost.

33 During one year a car travelled 7500 km. It used 1000 litres of petrol costing 8·3p per litre and 8 litres of oil costing 27·5p per litre. The road tax cost £25 and insurance cost £17·50. Repairs cost £24·40. In addition the value of the car fell by £120. Calculate the total cost of owning and running the car during the year, and also the cost per km rounded off to the nearest penny.

34 A shop advertises a discount of 5% for all cash payments. How much (to the nearest penny) would be paid by a customer for purchases totalling:

   a £17   *b* £15·60   *c* £35·50   *d* £23·82?

35 One dozen articles cost £2·90. If they are sold for 25p each, calculate the profit as a percentage of the selling price.

Cumulative Revision Exercises 243

36 A Trustee Savings Bank pays interest of $2\frac{1}{2}\%$ per annum on ordinary accounts and $6\frac{1}{2}\%$ per annum on special accounts, both calculated on the whole number of £s in the account. How much interest is paid in one year to a person who had £56 in his ordinary account and £145·59 in his special account?

37 Calculate the cost of the following hotel bill in Paris:
7 days room and breakfast at 52·40 francs per day
5 lunches at 9·60 francs each
14 dinners at 15·60 francs each.
  If there is a service charge of $12\frac{1}{2}\%$ to be added to the total, what change is given from a 1000-franc note?

38 A space probe travelling at an average speed of 15000 km/h took 240 days to reach the planet Venus. How many km did it travel? Express your answer in standard form.

39 The distance from Prestwick to New York is 5200 km. On the westward journey an aircraft flew at an average speed of 900 km/h and on the return journey at 1000 km/h. Find the difference in the times taken, to the nearest minute.

40 A number is chosen at random from the sequence 2, 3, 5, 8, 12, 17. Find the probability that it is: *a* even *b* odd *c* prime *d* odd and prime *e* odd or prime or both.

41 Assuming all days of the year are equally likely, how many pupils in a school of 1095 pupils are likely to have a birthday: *a* in December *b* on Christmas day?

42 Write out an array showing all possible ways of selecting a number at random from the set {1, 2, 3} and then selecting a second number from the set {3, 4, 5}. Use it to find the probability that the sum of the two numbers is: *a* equal to 6 *b* greater than 5 *c* equal to 9 *d* greater than 3.

43 Calculate the circumference and area of a circle which has:
a a radius of 14 cm *b* a diameter of 6 m.

44 From a square sheet of polyvinyl of side 35 cm are cut a square of side 13 cm, a rectangle 10·2 cm by 7·8 cm and a circle of radius 5·0 cm. Calculate the area of polyvinyl left, taking 3·14 as an approximation for π.

Arithmetic

45 Here is the timetable for the night sleeper trains from King's Cross to the North:

King's Cross	22 15	23 10	23 35
Peterborough	—	00 38	—
Edinburgh	—	06 45	06 47
Aberdeen	09 07	10 44	10 44

a How long does each train take from King's Cross to Aberdeen?
b How long does the 23 10 from King's Cross take to travel to Peterborough?
c Given that the distance from King's Cross to Edinburgh is 630 km, find the average speed of the slower train between King's Cross and Edinburgh to the nearest unit.

46 The petrol tank of a car holds 50 litres of petrol. The car can travel 9·6 km on a litre of petrol. How far can it travel on a full tank without filling up? If petrol costs 7·2p per litre, calculate as a decimal of a penny the average cost of petrol per kilometre.

47 A city savings bank, in conjunction with 90 school groups, carried through 153 630 transactions in one year, involving £26 668. Calculate: *a* the average number of transactions per group *b* the average sum of money involved in each transaction to the nearest penny.

48 A firm employs 12 skilled workers each earning £26 per week, and 8 unskilled men each earning £17 per week. Calculate the average wage earned per week by the firm's employees.

49 16 girls enter for a singles knock-out tennis championship. How many matches does the winner have to play? How many matches are there altogether?

50 A dog is chasing a rabbit. The rabbit has a start of 150 feet. The dog jumps 9 feet every time the rabbit jumps 7 feet. In how many jumps does the dog catch the rabbit? (Alcuin of York; about A.D. 800.)

Exercise B

1 Find all the prime factors of 504 and of 540. Hence find all the prime factors which are common to 504 and 540. Find also the smallest number (expressed in prime factors) into which both 504 and 540 will divide exactly (i.e. the LCM).

Cumulative Revision Exercises 245

2. The table shows the results of the binary operation ◇ ('diamond') on the set {0, 2, 4, 6, 8}

◇	0	2	4	6	8
0	0	0	0	0	0
2	0	4	8	2	6
4	0	8	6	4	2
6	0	2	4	6	8
8	0	6	2	8	4

(First number on left, Second number on top)

- *a* Is the table symmetrical about the leading diagonal? What does this tell you about the operation ◇?
- *b* Choose any replacements for *a*, *b* and *c* from the set {0, 2, 4, 6, 8} and find out if $(a \diamond b) \diamond c = a \diamond (b \diamond c)$. Repeat using different replacements. What law appears to be true?
- *c* Is there an identity element? (that is, is there an element I of the set such that $I \diamond a = a \diamond I = a$ for every element *a* of the set?) If so, which element is I?

3. Arrange in order of size, using the symbol $<$:

- *a* $\frac{8}{31}, \frac{7}{33}, \frac{7}{31}$
- *b* $\frac{3}{10}, \frac{1}{3}, \frac{4}{15}$

By expressing each fraction as a decimal, arrange in order of size, using the symbol $>$:   *c* $\frac{49}{60}, \frac{53}{64}, \frac{25}{31}$

4. One way to simplify $\frac{\frac{7}{8}}{1\frac{1}{2}}$ is to multiply numerator and denominator both by 8, obtaining $\frac{7}{12}$. In the same way a 'complex' fraction like $\frac{\frac{1}{2}+\frac{1}{3}}{\frac{1}{4}-\frac{1}{6}}$ can be simplified by multiplying each term in the numerator and denominator by 12. Why do we choose 12? Simplify the fraction.

Now simplify: *a* $\dfrac{\frac{3}{4}-\frac{1}{6}}{\frac{5}{6}+\frac{2}{3}}$ *b* $\dfrac{\frac{3}{5}-\frac{1}{4}}{\frac{7}{10}+\frac{1}{2}}$ *c* $\dfrac{\frac{2}{3}+\frac{3}{5}}{\frac{1}{6}+\frac{2}{15}}$

5. Express in standard form, rounded off to three significant figures:
- *a* 300 000 000
- *b* 29·1 million
- *c* 79·26
- *d* 4998

6. The marks scored in a skating contest were 27·75, 29·50, 22·25, 34·50, 49·25, 51·75, 46·00, 41·50. Calculate the average score.

7. How many pieces of wire each 7·8 cm long can be cut from a coil containing 10 m of wire? What length will be left over?

Arithmetic

8. The cost of setting up the type for a school magazine is £65. The cost of printing is £5·50 per 100 copies. The cost of paper, ink, etc., is 6p per copy. 700 copies are printed. If advertisements draw in £86, what is the minimum number of copies that must be sold at 10p each to avoid a loss on the venture?

9. A rectangular plot 13·3 m long and 6·5 m broad is to be made into a lawn. Grass seed is to be sown at the rate of 60 g/m², and can be bought in ½-kg bags. How many bags should be bought?

10. The perimeter of a rectangle is 28 mm, and the ratio of the length to the breadth is 4:3. Find its area.

11. A rectangular water tank has a base 0·75 m by 0·24 m, and holds 63 litres of water. Calculate the depth of the water.

12. What is the greatest number of rectangular packets each 9 cm by 7 cm by 6 cm which can be packed, all lying the same way, into a rectangular case 1 m long, 40 cm broad and 80 cm high?

13. Which of the following are approximations for $\sqrt{0·745}$?
    a $8·63 \times 10^{-1}$ *b* $2·73 \times 10^{-1}$ *c* $0·863$ *d* $0·273$

14. Express the square root of $3·45 \times 10^7$ in standard form.

15. Use tables to find $\sqrt{7·86}$ to 3 significant figures. Use this result as a first estimate to calculate the square root to 5 significant figures, doing one division. Hence find $\sqrt{0·000786}$ to 5 significant figures.

16. One cubic metre of copper is drawn into a uniform wire 500 km long. Find the area of cross-section of the wire as a decimal fraction of a square centimetre.

17. The cost of spraying a field of area ¾ hectare is £4·20. What is the cost of spraying a rectangular field measuring 200 m by 160 m? (1 hectare = 10 000 m²)

18. A slow train averaging 48 km/h takes 3½ hours for a journey. How long would an express averaging 90 km/h take? What is the length of the journey?

19. A car is 3·80 m long and 1·68 m wide. If a model of the car is to be 19 cm long, how wide should it be?

20. When planting strawberry plants, I require 6 rows if I place the plants 36 cm apart. How many rows would I require if I set the plants 30 cm apart?

Cumulative Revision Exercises

21. A boy answered the first six questions in a mathematics test in 20 minutes. How long will it take him to finish the paper, which contains ten questions?

22. A plan is drawn to a scale of 1:20000. What distance is represented by a line 1 cm long on the map? What area of the ground is represented by 1 cm^2 on the map? What area is represented by a square of side 1·8 cm? (10000 m^2 = 1 hectare)

23. If 1·75 litres of a gas weigh 0·630 g, what is the mass of 22·5 litres of this gas?

24. Write down all the pairs of numbers in binary form whose sum is 100_{two}.

25. Calculate: *a* the perimeter of a rectangle of length 11011_{two} metres and breadth 1010_{two} metres   *b* the length of a side of a square of perimeter 10100_{two} metres.

26. Arrange in order of size, using the symbol <:

    100110, 101010, 101100, 110001, 100011.

27. Calculate the volume of a cuboid of dimensions 101 cm, 110 cm and 11 cm, the numbers being in the scale of two.

28. In the scale of ten we know that $0·1 = \dfrac{1}{10}$ and $0·01 = \dfrac{1}{100} = \dfrac{1}{10 \times 10}$. What then must be the meanings of 0·1 and 0·01 in the scale of two? Express $0·11_{two}$ and $0·101_{two}$ in the same way, and then convert these to decimal form. ($0·1_{two}$, etc. are sometimes called 'bicimals'.)

29. In one year a householder used 3600 litres of oil for central heating. Oil cost 1·455p per litre, and annual servicing of the boiler cost £4·50. Find the total annual cost, and the monthly payment necessary to meet this cost.

30. A box of 48 packets of potato crisps is bought for 75p and the packets are sold at 2½p each. Find the profit as a percentage: *a* of the cost price   *b* of the selling price.

31. A jeweller buys a brooch at a wholesale price of £23·25. To calculate his selling price he has to add $66\frac{2}{3}\%$ of the wholesale price to cover tax, and a further 40% of the wholesale price to cover his own expenses and profit. What is his selling price?

Arithmetic

32. In each of the following, calculate what percentage the first is of the second: *a* 37; 150   *b* £5·50; £73·20   *c* 2·25 cm; 1·5 m.

33. Calculate the simple interest on £112·50 at $3\frac{1}{2}\%$ per annum from 20th January to 3rd July 1971, interest being payable on complete pounds and for complete calendar months.

34. A man has £550 to invest and he is considering two alternatives:
    Either he can invest it all in a Building Society which will pay $5\frac{1}{2}\%$ per annum interest;
    Or he can invest £100 in a Savings Bank Ordinary Account which pays $2\frac{1}{2}\%$ p.a. interest, and the rest in a Special Investment account which pays $6\frac{1}{2}\%$ p.a. interest.
    Which alternative gives the better return per year, and by how much?

35. In a certain year a manufacturer sold 24 000 transistor radios at £5·50 each. The following year he reduced the price of each radio by 20% and as a result the number of radios sold rose by $37\frac{1}{2}\%$. Find the increase in the value of his sales as a percentage of the total value of his sales in the first year.

36. A box contains one red, one yellow, one blue and one white disc. If a disc is selected at random, what is the probability that:
    a it is blue   *b* it is not blue?
    The disc is replaced and again a disc is drawn. By drawing up an array of all possible outcomes, find the probability that a red and a blue disc have been selected (not necessarily in that order) in two successive draws.

37. A number is selected from the set {1, 3, 5, 7} and added to a number selected from the set {2, 4, 6, 8}; show all the possible sums in an array. From this draw up a frequency table showing how often each sum occurs. Calculate the most frequent sum and the probability that it occurs.
    Calculate also the probability that the sum is:
    a 13   *b* odd   *c* even   *d* divisible by 3

38. In a certain country the number of men in the 55-64 age group is 3·456 million. On the basis of past statistics it is estimated that the probability that any one man in this group will be alive ten years later is 0·68; estimate the number of men in the 65-74 age group ten years later.

Cumulative Revision Exercises 249

39 Of 80 pupils in a certain year, 20 are specializing in mathematics and 30 are members of the athletic club; 40 of the pupils neither specialize in mathematics nor are members of the athletic club. Show these facts in a Venn diagram. Hence find the probability that if a pupil is selected at random: *a* he both specializes in mathematics and is a member of the athletic club *b* he specializes in mathematics and is not a member of the athletic club.

40 The daily BOAC plane leaves Glasgow Prestwick at 1250 and arrives in New York at 1505. The return flight leaves New York at 1930 and arrives at Glasgow at 0655. These are local times; travelling west a passenger must put his watch back 5 hours; travelling East he must put his watch forward 5 hours. If the distance is 5200 km, find the average speed on each flight.

41 A man lives 56 km from Inverness. He motored there in daylight at an average speed of 60 km/h and returned after dark at an average speed of 40 km/h. Calculate his average speed for the whole journey.

42 The average mass of 8 oarsmen in a boat-race crew is 91·6 kg. The average mass of the 9 crew members, including the cox, is 89·8 kg. What is the mass of the cox?

43 The sample census of 1966 in Scotland showed that there were 1600000 households in Scotland totalling 4987000 persons. Find the average number of persons per household, rounded off to 2 significant figures.

Out of the whole population, 14% were of pensionable age and of these 209000 were men. How many women of pensionable age were there, rounded off to 3 significant figures?

44a Calculate $(2\frac{1}{3} - 1\frac{1}{2}) \div 6\frac{2}{3}$
 b Express the product of $4·5 \times 10^3$ and $3·7 \times 10^5$ in the form $a \times 10^n$ with a rounded off to two significant figures.

45 A bicycle wheel has a diameter of 66 cm. How many times will it revolve in travelling 1 km without slipping?

46 A uniform circular disc of radius 5 cm weighs 300 g. Find its mass when $\frac{1}{2}$ cm is cut off all round the edge.

Arithmetic

47 Figure 3 shows a plan of a football field which consists of a rectangle 100 m by 60 m with a semicircle drawn externally on each end.

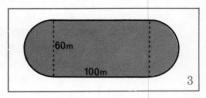

Calculate: *a* the perimeter *b* the area. If the plan is drawn to a scale of 1:1000, *write down* the dimensions, the perimeter and the area of the plan.

48 Oxford beat Cambridge in the 1966 Boat Race by $3\frac{3}{4}$ lengths. If one length was rowed in $3\frac{1}{2}$ seconds and the winner's time was 19 minutes 12 seconds, what was the loser's time to the nearest second?

49 A chemical mixture consists of three elements A, B and C in the ratio 4:3:2 by mass. 24 g of B and 10 g of C are in stock. What masses of A and C should be added to the quantities in stock of B and C to make the mixture, and what mass of mixture is obtained?

50 A merchant visited 3 fairs. At the first he doubled his money and then spent 30 crowns; at the second he tripled his money and then spent 54 crowns; at the third he quadrupled his money and then spent 72 crowns. He then had 48 crowns left. How much did he start with? (From a textbook dated 1484)

True-False Revision Exercise

State whether each of the following is true (T) or false (F).

1 $11011_{two} = 29_{ten}$

2 $3.75^2 = 14.1$, by slide rule.

3 $\sqrt{47.0} = 21.7$ to three significant figures.

4 $1\frac{1}{2}\%$ of £2300 = £34.

5 $4\frac{1}{2}$ dozen wallflowers at 17p per dozen cost $76\frac{1}{2}$p.

6 The length, breadth, height, and volume of a cuboid are 0·8 cm, 0·6 cm, 0·3 cm, and 1·44 cm³ respectively.

7 Cycling at 8 km/h I take 36 minutes for a journey, so cycling at 12 km/h I should take 24 minutes.

8 The circumference of a circle of radius 2 m is approximately 6·28 m.

Cumulative Revision Exercises

9 The probability that a number selected at random from the set $\{1, 2, 3, 4, 5\}$ is greater than 3 is 0·6.

10 Travelling for $2\frac{1}{2}$ hours at an average speed of 46 km/h, I should cover 115 km.

11 If the radius of a circle is doubled, its area will be doubled.

12 A rectangle 1·2 m by 98 cm has an area greater than 1 m².

13a $\sqrt{3\frac{1}{16}} = 1\frac{3}{4}$ b $\sqrt{0·5} \doteq 0·7$ c $\sqrt{239}$ lies between 15 and 16

14a $52·3 \times 0·03 = 15·69$ b $9·43 \div 0·471 \doteq 2$
 c $7·6 \times 10^4 = 76\,000$
 d If $a = 0·8$ and $b = 0·02$, $a + \frac{5}{2}b = \frac{17}{20}$

15 If a train travelled for 3 h 15 min at an average speed of 48 km/h,
 a the distance travelled was 156 km;
 b if the train stopped in stations for 15 min, the average running speed was 60 km/h;
 c if the length of the journey was doubled, and the speed halved, the time would be four times as long.

Computer Studies

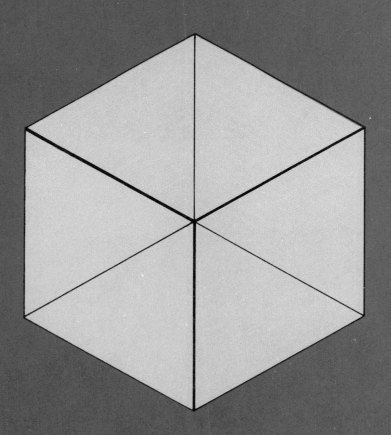

Computer Studies

Note to the Teacher on Chapter 1

In daily life we are all faced with problems for which we have devised our own method of solution or *algorithm*. In mathematics we are faced with a similar situation where we have to devise algorithms to solve problems. The fundamental idea of an algorithm is simply an ordered set of instructions. Algorithms can be described by flow charts, and consequently some skill must be developed in flow charting. It is a waste of time to give too much attention to linear flow charts since most real-life problems involve branching and looping.

When using a computer to solve a problem it must be realized that programming is only one step in the solution of the problem. The complete sequence of events is as follows:

  (i) Analysis of the problem
 (ii) Preparation of flow charts
(iii) Preparation of a program
 (iv) Data preparation
  (v) Testing and running the program
 (vi) Results

The most important stages in the solution of a problem are the first two, which define the problem and define the solution, respectively. Programming merely expresses the solution in a form suitable for computer interpretation.

Pupils should be encouraged therefore to draw flow charts which define the solution to the problem, and to use these as precise plans for producing the set of program instructions. This approach ensures that when writing a program pupils are concentrating solely on the problems of language, and are not sidetracked by the logic of the solution.

Flow Charts

1 Constructing a flow chart

In industry, commerce, government, education, science, medicine and many other spheres of modern life, computers are valuable servants of mankind. Most of the computers in use today are used for storing and processing information, but some are employed for performing high-speed calculations.

To do either of these jobs the computer must be 'programmed' with a sequence of simple instructions.

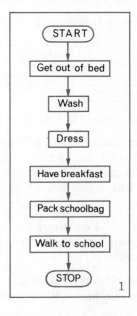

1

Figure 1 shows a possible sequence of instructions for the daily routine of getting ready for school, written out in the form of a *flow chart*.

Computer Studies

Certain conventions are used in flow charts:

(i) The words START and STOP are enclosed in *terminator* boxes shaped like this: ⊂⎯⎯⎯⎯⎯⎯⊃

(ii) Instructions are enclosed in *instruction* or *operation* boxes shaped like this: ☐

(iii) *Arrows* join up the instructions and show the order in which they are to be carried out.

Flow charts assist in the preparation of computer programs, which can be fed into the computer in the form of punched cards or paper tape as we saw in Book 2.

Exercise 1

1. Rearrange the instructions in Figure 1 in another *possible* sequence, and draw the corresponding flow chart.

2. Here are jumbled up instructions for *telephoning a friend*:

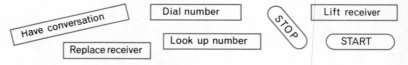

From these draw a suitable flow chart.

In each of the questions *3-6*, a familiar activity has been broken up into a set of simple instructions. Put the instructions into a suitable order and make a flow chart, using boxes of the proper shapes and inserting all necessary arrows. Remember to put in START and STOP boxes.

3. *Making a pot of tea.* **Pour water into teapot. Put tea in teapot. Fill electric kettle. Wait for kettle to boil. Switch off kettle. Switch on kettle. Plug in kettle.**

4. *Washing your hands.* **Take out plug. Turn on water. Turn off water. Put in plug. Dry your hands. Rinse your hands. Soap your hands.**

5. *Cleaning shoes.* **Polish with cloth. Scrape off mud. Polish with brush. Wash and dry your hands. Put on polish. Brush off mud. Put away shoe-cleaning things. Get out shoe-cleaning things.**

Constructing a flow chart

6 *Solving the equation* $12x - 15 = 9$. **Multiply each side by $\frac{1}{12}$. Add 15 to each side. Give the solution set. Write down the equation.**

7 Draw a flow chart for solving the equation $10x + 9 = x + 27$, and use it to solve the equation.

8 Make a flow chart for the following calculation, for which the instructions are given in order. **Think of a number between 100 and 500. Write down the number with its digits in reverse order. Calculate their difference. Divide the difference by 11. Write down the remainder.**

Use your flow chart for three different numbers. Can you explain the result?

Exercise 1B

1 Draw flow charts giving instructions to a person going from:
   a your mathematics room to the school office
   b the school office to your mathematics room.
   Compare the two flow charts.

2 Make a flow chart for the following instructions, given in order. **Think of a number. Double it. Add 7. Multiply by 5. Add 15. Divide by 10. Subtract the number you first thought of.**

Use your flow chart for three different numbers. Can you explain the result?

3 Make a flow chart for converting marks out of 150 to percentages.

4 (1, 3), (2, 5), (5, 11) are ordered pairs given by the mapping $x \to 2x + 1$. Make a flow chart which gives the second element of such an ordered pair when the first is known. Use the flow chart to complete the ordered pairs: (3,), (0,), (−2,).

5 Draw a flow chart for the mapping $x \to 3x - 1$, and use it to complete the ordered pairs: (1,), (2,), (5,).

6 p and q are natural numbers. M_p is the set of multiples of p up to $p \times q$, and M_q is the set of multiples of q up to $p \times q$.

Make a flow chart for finding the LCM of p and q, using the following instructions. **List M_p. List M_q. List $M_p \cap M_q$. Write down the least member of $M_p \cap M_q$** after zero, giving the LCM of p and q. Use your flow chart to find the LCM of 3 and 4; and of 6 and 8.

Computer Studies

7 Give instructions in the form of a flow chart for using the Sieve of Eratosthenes to find all prime numbers less than 100. Do not forget STOP at the end.

8 Make flow charts to explain precisely how to use a slide rule:
  a to find the quotient of two numbers, using the C and D scales
  b to find the square root of a number, using the A and D scales.

2 Decision boxes

Figure 2 shows a flow chart for finding whether a number is even or odd. Notice that the second *instruction* may be either **Write 'even'** or **Write 'odd'**. The flow chart allows for the two cases, *even* and *odd*. In programming a computer it is essential to allow for every case.

2

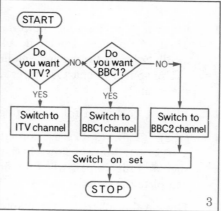
3

In order to allow for different possibilities, *decision boxes* like the diamond-shaped box in Figure 2 are used. If there are more than two possibilities, more than one decision box is needed.

Figure 3 shows instructions for switching on television, assuming that the set is plugged in and in good order, and that BBC1 or BBC2 or ITV is to be chosen.

Notice that only two flow lines come from each decision box.

Decision boxes

This corresponds to the working of the computer, and to the True-False nature of everyday logic.

The question in the decision box must be worded in such a way that the answers are YES or NO, as shown in Figures 2 and 3. It would not have been suitable to ask the question 'Do you want BBC1 or BBC2?' in the second decision box in Figure 3. Why not?

We must always make sure that each route leads eventually to the STOP box.

Exercise 2

1. Put the following instructions for boiling an egg into a suitable order, and draw a flow chart. **Remove egg from pan. Place pan on cooker. Wash the pan. Switch off cooker. Switch on cooker. Is the pan clean? Place egg in pan when water boils. Put water in pan. Leave egg in boiling water for three minutes.**

2. Your friend is thirsty and wishes to have something to drink. You have tea, coffee and lemonade to offer him. Make a suitable flow chart from the following:

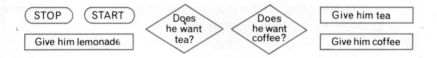

3. Make a flow chart from the following instructions to find whether or not a number is divisible by 30, by finding whether or not it is divisible by 2, 3 and 5. **Write down number. Is number divisible by 2? Number is divisible by 30. Is number divisible by 5? Number is not divisible by 30. Is number divisible by 3?**

4. Draw a flow chart for making a bread-and-butter sandwich with banana, cheese or jam. It is known that jam is available, but the eater likes banana best, and prefers cheese to jam.

5. Figure 4 shows how to find if a year is a leap year.
    a. Use it to test: (*1*) 1976 (*2*) 1984 (*3*) 2000 (*4*) 2001 (*5*) 2100
    b. From the flow chart give as briefly as you can the meaning of 'leap year'.

Computer Studies

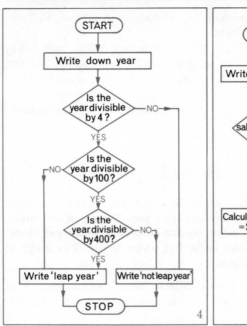

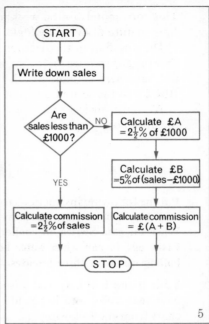

6   Figure 5 shows how to calculate an agent's commission based on his weekly sales. Calculate the commission payable in three successive weeks when the value of goods sold was:

    a   £720   b   £2000   c   £1800.

7 a   Make a flow chart for calculating the selling price of goods during a sale. On goods marked at less than £10 the discount is 5%; on goods marked £10 or more the discount is 10%. You may find some of the following instructions helpful (also Figure 5). **Calculate discount = 5% of marked price. Calculate discount = *10*% of marked price. Calculate sale price = marked price − discount.**

   b   Use your flow chart to find the sale prices of goods marked:
      (*1*) £5   (*2*) £15   (*3*) £10.

8 a   Make a flow chart from the following to find the product of two numbers using a slide rule. **Set cursor against first number on D scale. Set 1 on C scale against cursor. Can you read on D scale against second number on C scale? Set 10 on C scale against cursor. Read on D scale against second number on C scale.**

   b   Use your flow chart to multiply by slide rule:
      (*1*) 3×4 (*2*) $2 \cdot 57 \times 3 \cdot 21$

3 Loops

An instruction like 'Wait until the kettle boils' is too complex for a flow chart. It really consists of both questions and instructions; Figure 6 shows how it could be broken down into these parts. Notice how the same sequence of 'instruction—then—question' occurs again and again. To prevent having to write down the sequence again and again in the flow chart we introduce a *loop* as shown in Figure 7.

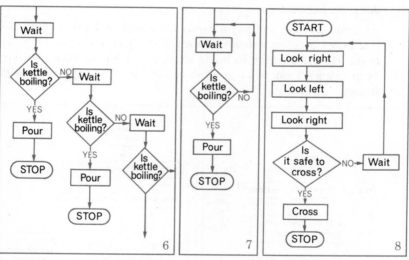

Figure 8 shows a simplified version of Kerb Drill in the form of a flow chart. Study the various boxes and routes in the diagram, and note particularly the use of the loop.

Exercise 3

Draw flow charts for the instructions given in questions *1-5*, introducing loops where necessary.

1   *Washing your hands.* **Is there a towel available? Turn on hot tap. Get cake of soap. Are your hands clean now? Rinse off soap. Get towel. Is there any soap? Wash hands with soap. Turn off water. Wet your hands. Dry your hands.**

2   *Cleaning a pair of shoes.* **Put cleaning-things away. Are your hands**

dirty? Polish shoes with cloth. Put polish on shoes. Get cleaning-things. Are shoes clean? Scrape off mud. Wash your hands. Is the mud off shoes? Get a scraper. Polish shoes with brush. Is there mud on shoes?

3   *Starting with 50, add 10 repeatedly until the sum is 100.* **Add 10. Write down 50. Write down 100. Is the sum 100?**

4   *Starting with 512, divide by 2 repeatedly until the quotient is 1.* **Divide by 2. Write down 1. Write down 512. Is the quotient 1?**

5   *Starting with 40, subtract 3 repeatedly until the difference is less than zero.* **Is the difference negative? Write down the difference. Write down 40. Subtract 3.**

6   At a coconut shy you pay a penny for a ball, and win a goldfish if you knock a coconut down. Draw a flow chart, including the questions: Have I a penny to buy a shot? Did I hit the coconut? You keep trying till you win a goldfish or run out of money, then you go home.

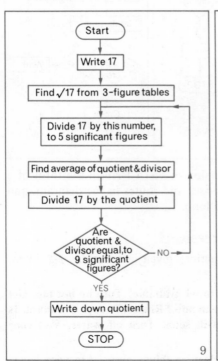

9

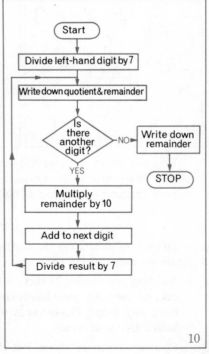

10

Loops

7   Figure 9 gives a flow chart for finding the square root of 17 to 9 significant figures, using an iterative method.
Use the flow chart to carry out the calculation.

8   Figure 10 shows a flow chart for short division by 7. Use it to find the quotient and remainder when the following are divided by 7:

   a 12   b 88   c 100   d 651   e 3   f 5

9   Make a flow chart for expressing a number in binary form, according to the following instructions, and inserting the necessary loop. **Write down the number. Divide the number by 2. Write down the quotient and remainder. Is the quotient zero? Arrange remainders in order, the last one on the left, to obtain the binary form of the original number.**

Squares from 1 to 10

	0	1	2	3	4	5	6	7	8	9
1·0	1·00	1·02	1·04	1·06	1·08	1·10	1·12	1·14	1·17	1·19
1·1	1·21	1·23	1·25	1·28	1·30	1·32	1·35	1·37	1·39	1·42
1·2	1·44	1·46	1·49	1·51	1·54	1·56	1·59	1·61	1·64	1·66
1·3	1·69	1·72	1·74	1·77	1·80	1·82	1·85	1·88	1·90	1·93
1·4	1·96	1·99	2·02	2·04	2·07	2·10	2·13	2·16	2·19	2·22
1·5	2·25	2·28	2·31	2·34	2·37	2·40	2·43	2·46	2·50	2·53
1·6	2·56	2·59	2·62	2·66	2·69	2·72	2·76	2·79	2·82	2·86
1·7	2·89	2·92	2·96	2·99	3·03	3·06	3·10	3·13	3·17	3·20
1·8	3·24	3·28	3·31	3·35	3·39	3·42	3·46	3·50	3·53	3·57
1·9	3·61	3·65	3·69	3·72	3·76	3·80	3·84	3·88	3·92	3·96
2·0	4·00	4·04	4·08	4·12	4·16	4·20	4·24	4·28	4·33	4·37
2·1	4·41	4·45	4·49	4·54	4·58	4·62	4·67	4·71	4·75	4·80
2·2	4·84	4·88	4·93	4·97	5·02	5·06	5·11	5·15	5·20	5·24
2·3	5·29	5·34	5·38	5·43	5·48	5·52	5·57	5·62	5·66	5·71
2·4	5·76	5·81	5·86	5·90	5·95	6·00	6·05	6·10	6·15	6·20
2·5	6·25	6·30	6·35	6·40	6·45	6·50	6·55	6·60	6·66	6·71
2·6	6·76	6·81	6·86	6·92	6·97	7·02	7·08	7·13	7·18	7·24
2·7	7·29	7·34	7·40	7·45	7·51	7·56	7·62	7·67	7·73	7·78
2·8	7·84	7·90	7·95	8·01	8·07	8·12	8·18	8·24	8·29	8·35
2·9	8·41	8·47	8·53	8·58	8·64	8·70	8·76	8·82	8·88	8·94
3·0	9·00	9·06	9·12	9·18	9·24	9·30	9·36	9·42	9·49	9·55
3·1	9·61	9·67	9·73	9·80	9·86	9·92	9·99	10·05	10·11	10·18
3·2	10·24	10·30	10·37	10·43	10·50	10·56	10·63	10·69	10·76	10·82
3·3	10·89	10·96	11·02	11·09	11·16	11·22	11·29	11·36	11·42	11·49
3·4	11·56	11·63	11·70	11·76	11·83	11·90	11·97	12·04	12·11	12·18
3·5	12·25	12·32	12·39	12·46	12·53	12·60	12·67	12·74	12·82	12·89
3·6	12·96	13·03	13·10	13·18	13·25	13·32	13·40	13·47	13·54	13·62
3·7	13·69	13·76	13·84	13·91	13·99	14·06	14·14	14·21	14·29	14·36
3·8	14·44	14·52	14·59	14·67	14·75	14·82	14·90	14·98	15·05	15·13
3·9	15·21	15·29	15·37	15·44	15·52	15·60	15·68	15·76	15·84	15·92
4·0	16·00	16·08	16·16	16·24	16·32	16·40	16·48	16·56	16·65	16·73
4·1	16·81	16·89	16·97	17·06	17·14	17·22	17·31	17·39	17·47	17·56
4·2	17·64	17·72	17·81	17·89	17·98	18·06	18·15	18·23	18·32	18·40
4·3	18·49	18·58	18·66	18·75	18·84	18·92	19·01	19·10	19·18	19·27
4·4	19·36	19·45	19·54	19·62	19·71	19·80	19·89	19·98	20·07	20·16
4·5	20·25	20·34	20·43	20·52	20·61	20·70	20·79	20·88	20·98	21·07
4·6	21·16	21·25	21·34	21·44	21·53	21·62	21·72	21·81	21·90	22·00
4·7	22·09	22·18	22·28	22·37	22·47	22·56	22·66	22·75	22·85	22·94
4·8	23·04	23·14	23·23	23·33	23·43	23·52	23·62	23·72	23·81	23·91
4·9	24·01	24·11	24·21	24·30	24·40	24·50	24·60	24·70	24·80	24·90
5·0	25·00	25·10	25·20	25·30	25·40	25·50	25·60	25·70	25·81	25·91
5·1	26·01	26·11	26·21	26·32	26·42	26·52	26·63	26·73	26·83	26·94
5·2	27·04	27·14	27·25	27·35	27·46	27·56	27·67	27·77	27·88	27·98
5·3	28·09	28·20	28·30	28·41	28·52	28·62	28·73	28·84	28·94	29·05
5·4	29·16	29·27	29·38	29·48	29·59	29·70	29·81	29·92	30·03	30·14

Squares from 1 to 10

	0	1	2	3	4	5	6	7	8	9
5·5	30·25	30·36	30·47	30·58	30·69	30·80	30·91	31·02	31·14	31·25
5·6	31·36	31·47	31·58	31·70	31·81	31·92	32·04	32·15	32·26	32·38
5·7	32·49	32·60	32·72	32·83	32·95	33·06	33·18	33·29	33·41	33·52
5·8	33·64	33·76	33·87	33·99	34·11	34·22	34·34	34·46	34·57	34·69
5·9	34·81	34·93	35·05	35·16	35·28	35·40	35·52	35·64	35·76	35·88
6·0	36·00	36·12	36·24	36·36	36·48	36·60	36·72	36·84	36·97	37·09
6·1	37·21	37·33	37·45	37·58	37·70	37·82	37·95	38·07	38·19	38·32
6·2	38·44	38·56	38·69	38·81	38·94	39·06	39·19	39·31	39·44	39·56
6·3	39·69	39·82	39·94	40·07	40·20	40·32	40·45	40·58	40·70	40·83
6·4	40·96	41·09	41·22	41·34	41·47	41·60	41·73	41·86	41·99	42·12
6·5	42·25	42·38	42·51	42·64	42·77	42·90	43·03	43·16	43·30	43·43
6·6	43·56	43·69	43·82	43·96	44·09	44·22	44·36	44·49	44·62	44·76
6·7	44·89	45·02	45·16	45·29	45·43	45·56	45·70	45·83	45·97	46·10
6·8	46·24	46·38	46·51	46·65	46·79	46·92	47·06	47·20	47·33	47·47
6·9	47·61	47·75	47·89	48·02	48·16	48·30	48·44	48·58	48·72	48·86
7·0	49·00	49·14	49·28	49·42	49·56	49·70	49·84	49·98	50·13	50·27
7·1	50·41	50·55	50·69	50·84	50·98	51·12	51·27	51·41	51·55	51·70
7·2	51·84	51·98	52·13	52·27	52·42	52·56	52·71	52·85	53·00	53·14
7·3	53·29	53·44	53·58	53·73	53·88	54·02	54·17	54·32	54·46	54·61
7·4	54·76	54·91	55·06	55·20	55·35	55·50	55·65	55·80	55·95	56·10
7·5	56·25	56·40	56·55	56·70	56·85	57·00	57·15	57·30	57·46	57·61
7·6	57·76	57·91	58·06	58·22	58·37	58·52	58·68	58·83	58·98	59·14
7·7	59·29	59·44	59·60	59·75	59·91	60·06	60·22	60·37	60·53	60·68
7·8	60·84	61·00	61·15	61·31	61·47	61·62	61·78	61·94	62·09	62·25
7·9	62·41	62·57	62·73	62·88	63·04	63·20	63·36	63·52	63·68	63·84
8·0	64·00	64·16	64·32	64·48	64·64	64·80	64·96	65·12	65·29	65·45
8·1	65·61	65·77	65·93	66·10	66·26	66·42	66·59	66·75	66·91	67·08
8·2	67·24	67·40	67·57	67·73	67·90	68·06	68·23	68·39	68·56	68·72
8·3	68·89	69·06	69·22	69·39	69·56	69·72	69·89	70·06	70·22	70·39
8·4	70·56	70·73	70·90	71·06	71·23	71·40	71·57	71·74	71·91	72·08
8·5	72·25	72·42	72·59	72·76	72·93	73·10	73·27	73·44	73·62	73·79
8·6	73·96	74·13	74·30	74·48	74·65	74·82	75·00	75·17	75·34	75·52
8·7	75·69	75·86	76·04	76·21	76·39	76·56	76·74	76·91	77·09	77·26
8·8	77·44	77·62	77·79	77·97	78·15	78·32	78·50	78·68	78·85	79·03
8·9	79·21	79·39	79·57	79·74	79·92	80·10	80·28	80·46	80·64	80·82
9·0	81·00	81·18	81·36	81·54	81·72	81·90	82·08	82·26	82·45	82·63
9·1	82·81	82·99	83·17	83·36	83·54	83·72	83·91	84·09	84·27	84·46
9·2	84·64	84·82	85·01	85·19	85·38	85·56	85·75	85·93	86·12	86·30
9·3	86·49	86·68	86·86	87·05	87·24	87·42	87·61	87·80	87·98	88·17
9·4	88·36	88·55	88·74	88·92	89·11	89·30	89·49	89·68	89·87	90·06
9·5	90·25	90·44	90·63	90·82	91·01	91·20	91·39	91·58	91·78	91·97
9·6	92·16	92·35	92·54	92·74	92·93	93·12	93·32	93·51	93·70	93·90
9·7	94·09	94·28	94·48	94·67	94·87	95·06	95·26	95·45	95·65	95·84
9·8	96·04	96·24	96·43	96·63	96·83	97·02	97·22	97·42	97·61	97·81
9·9	98·01	98·21	98·41	98·60	98·80	99·00	99·20	99·40	99·60	99·80

Square roots from 1 to 10

	0	1	2	3	4	5	6	7	8	9
1·0	1·00	1·00	1·01	1·01	1·02	1·02	1·03	1·03	1·04	1·04
1·1	1·05	1·05	1·06	1·06	1·07	1·07	1·08	1·08	1·09	1·09
1·2	1·10	1·10	1·10	1·11	1·11	1·12	1·12	1·13	1·13	1·14
1·3	1·14	1·14	1·15	1·15	1·16	1·16	1·17	1·17	1·17	1·18
1·4	1·18	1·19	1·19	1·20	1·20	1·20	1·21	1·21	1·22	1·22
1·5	1·22	1·23	1·23	1·24	1·24	1·24	1·25	1·25	1·26	1·26
1·6	1·26	1·27	1·27	1·28	1·28	1·28	1·29	1·29	1·30	1·30
1·7	1·30	1·31	1·31	1·32	1·32	1·32	1·33	1·33	1·33	1·34
1·8	1·34	1·35	1·35	1·35	1·36	1·36	1·36	1·37	1·37	1·37
1·9	1·38	1·38	1·39	1·39	1·39	1·40	1·40	1·40	1·41	1·41
2·0	1·41	1·42	1·42	1·42	1·43	1·43	1·44	1·44	1·44	1·45
2·1	1·45	1·45	1·46	1·46	1·46	1·47	1·47	1·47	1·48	1·48
2·2	1·48	1·49	1·49	1·49	1·50	1·50	1·50	1·51	1·51	1·51
2·3	1·52	1·52	1·52	1·53	1·53	1·53	1·54	1·54	1·54	1·55
2·4	1·55	1·55	1·56	1·56	1·56	1·57	1·57	1·57	1·57	1·58
2·5	1·58	1·58	1·59	1·59	1·59	1·60	1·60	1·60	1·61	1·61
2·6	1·61	1·62	1·62	1·62	1·62	1·63	1·63	1·63	1·64	1·64
2·7	1·64	1·65	1·65	1·65	1·66	1·66	1·66	1·66	1·67	1·67
2·8	1·67	1·68	1·68	1·68	1·69	1·69	1·69	1·69	1·70	1·70
2·9	1·70	1·71	1·71	1·71	1·71	1·72	1·72	1·72	1·73	1·73
3·0	1·73	1·73	1·74	1·74	1·74	1·75	1·75	1·75	1·75	1·76
3·1	1·76	1·76	1·77	1·77	1·77	1·77	1·78	1·78	1·78	1·79
3·2	1·79	1·79	1·79	1·80	1·80	1·80	1·81	1·81	1·81	1·81
3·3	1·82	1·82	1·82	1·82	1·83	1·83	1·83	1·84	1·84	1·84
3·4	1·84	1·85	1·85	1·85	1·85	1·86	1·86	1·86	1·87	1·87
3·5	1·87	1·87	1·88	1·88	1·88	1·88	1·89	1·89	1·89	1·89
3·6	1·90	1·90	1·90	1·91	1·91	1·91	1·91	1·92	1·92	1·92
3·7	1·92	1·93	1·93	1·93	1·93	1·94	1·94	1·94	1·94	1·95
3·8	1·95	1·95	1·95	1·96	1·96	1·96	1·96	1·97	1·97	1·97
3·9	1·97	1·98	1·98	1·98	1·98	1·99	1·99	1·99	1·99	2·00
4·0	2·00	2·00	2·00	2·01	2·01	2·01	2·01	2·02	2·02	2·02
4·1	2·02	2·03	2·03	2·03	2·03	2·04	2·04	2·04	2·04	2·05
4·2	2·05	2·05	2·05	2·06	2·06	2·06	2·06	2·07	2·07	2·07
4·3	2·07	2·08	2·08	2·08	2·08	2·09	2·09	2·09	2·09	2·10
4·4	2·10	2·10	2·10	2·10	2·11	2·11	2·11	2·11	2·12	2·12
4·5	2·12	2·12	2·13	2·13	2·13	2·13	2·14	2·14	2·14	2·14
4·6	2·14	2·15	2·15	2·15	2·15	2·16	2·16	2·16	2·16	2·17
4·7	2·17	2·17	2·17	2·17	2·18	2·18	2·18	2·18	2·19	2·19
4·8	2·19	2·19	2·20	2·20	2·20	2·20	2·20	2·21	2·21	2·21
4·9	2·21	2·22	2·22	2·22	2·22	2·22	2·23	2·23	2·23	2·23
5·0	2·24	2·24	2·24	2·24	2·24	2·25	2·25	2·25	2·25	2·26
5·1	2·26	2·26	2·26	2·26	2·27	2·27	2·27	2·27	2·28	2·28
5·2	2·28	2·28	2·28	2·29	2·29	2·29	2·29	2·30	2·30	2·30
5·3	2·30	2·30	2·31	2·31	2·31	2·31	2·32	2·32	2·32	2·32
5·4	2·32	2·33	2·33	2·33	2·33	2·33	2·34	2·34	2·34	2·34

Square roots from 1 to 10

	0	1	2	3	4	5	6	7	8	9
5·5	2·35	2·35	2·35	2·35	2·35	2·36	2·36	2·36	2·36	2·36
5·6	2·37	2·37	2·37	2·37	2·37	2·38	2·38	2·38	2·38	2·39
5·7	2·39	2·39	2·39	2·39	2·40	2·40	2·40	2·40	2·40	2·41
5·8	2·41	2·41	2·41	2·41	2·42	2·42	2·42	2·42	2·42	2·43
5·9	2·43	2·43	2·43	2·44	2·44	2·44	2·44	2·44	2·45	2·45
6·0	2·45	2·45	2·45	2·46	2·46	2·46	2·46	2·46	2·47	2·47
6·1	2·47	2·47	2·47	2·48	2·48	2·48	2·48	2·48	2·49	2·49
6·2	2·49	2·49	2·49	2·50	2·50	2·50	2·50	2·50	2·51	2·51
6·3	2·51	2·51	2·51	2·52	2·52	2·52	2·52	2·52	2·53	2·53
6·4	2·53	2·53	2·53	2·54	2·54	2·54	2·54	2·54	2·55	2·55
6·5	2·55	2·55	2·55	2·56	2·56	2·56	2·56	2·56	2·57	2·57
6·6	2·57	2·57	2·57	2·57	2·58	2·58	2·58	2·58	2·58	2·59
6·7	2·59	2·59	2·59	2·59	2·60	2·60	2·60	2·60	2·60	2·61
6·8	2·61	2·61	2·61	2·61	2·62	2·62	2·62	2·62	2·62	2·62
6·9	2·63	2·63	2·63	2·63	2·63	2·64	2·64	2·64	2·64	2·64
7·0	2·65	2·65	2·65	2·65	2·65	2·66	2·66	2·66	2·66	2·66
7·1	2·66	2·67	2·67	2·67	2·67	2·67	2·68	2·68	2·68	2·68
7·2	2·68	2·69	2·69	2·69	2·69	2·69	2·69	2·70	2·70	2·70
7·3	2·70	2·70	2·71	2·71	2·71	2·71	2·71	2·71	2·72	2·72
7·4	2·72	2·72	2·72	2·73	2·73	2·73	2·73	2·73	2·73	2·74
7·5	2·74	2·74	2·74	2·74	2·75	2·75	2·75	2·75	2·75	2·75
7·6	2·76	2·76	2·76	2·76	2·76	2·77	2·77	2·77	2·77	2·77
7·7	2·77	2·78	2·78	2·78	2·78	2·78	2·79	2·79	2·79	2·79
7·8	2·79	2·79	2·80	2·80	2·80	2·80	2·80	2·81	2·81	2·81
7·9	2·81	2·81	2·81	2·82	2·82	2·82	2·82	2·82	2·82	2·83
8·0	2·83	2·83	2·83	2·83	2·84	2·84	2·84	2·84	2·84	2·84
8·1	2·85	2·85	2·85	2·85	2·85	2·85	2·86	2·86	2·86	2·86
8·2	2·86	2·87	2·87	2·87	2·87	2·87	2·87	2·88	2·88	2·88
8·3	2·88	2·88	2·88	2·89	2·89	2·89	2·89	2·89	2·89	2·90
8·4	2·90	2·90	2·90	2·90	2·91	2·91	2·91	2·91	2·91	2·91
8·5	2·92	2·92	2·92	2·92	2·92	2·92	2·93	2·93	2·93	2·93
8·6	2·93	2·93	2·94	2·94	2·94	2·94	2·94	2·94	2·95	2·95
8·7	2·95	2·95	2·95	2·95	2·96	2·96	2·96	2·96	2·96	2·96
8·8	2·97	2·97	2·97	2·97	2·97	2·97	2·98	2·98	2·98	2·98
8·9	2·98	2·98	2·99	2·99	2·99	2·99	2·99	2·99	3·00	3·00
9·0	3·00	3·00	3·00	3·00	3·01	3·01	3·01	3·01	3·01	3·01
9·1	3·02	3·02	3·02	3·02	3·02	3·02	3·03	3·03	3·03	3·03
9·2	3·03	3·03	3·04	3·04	3·04	3·04	3·04	3·04	3·05	3·05
9·3	3·05	3·05	3·05	3·05	3·06	3·06	3·06	3·06	3·06	3·06
9·4	3·07	3·07	3·07	3·07	3·07	3·07	3·08	3·08	3·08	3·08
9·5	3·08	3·08	3·09	3·09	3·09	3·09	3·09	3·09	3·10	3·10
9·6	3·10	3·10	3·10	3·10	3·10	3·11	3·11	3·11	3·11	3·11
9·7	3·11	3·12	3·12	3·12	3·12	3·12	3·12	3·13	3·13	3·13
9·8	3·13	3·13	3·13	3·14	3·14	3·14	3·14	3·14	3·14	3·14
9·9	3·15	3·15	3·15	3·15	3·15	3·15	3·16	3·16	3·16	3·16

Square roots from 10 to 100

	·0	·1	·2	·3	·4	·5	·6	·7	·8	·9
10	3·16	3·18	3·19	3·21	3·22	3·24	3·26	3·27	3·29	3·30
11	3·32	3·33	3·35	3·36	3·38	3·39	3·41	3·42	3·44	3·45
12	3·46	3·48	3·49	3·51	3·52	3·54	3·55	3·56	3·58	3·59
13	3·61	3·62	3·63	3·65	3·66	3·67	3·69	3·70	3·71	3·73
14	3·74	3·75	3·77	3·78	3·79	3·81	3·82	3·83	3·85	3·86
15	3·87	3·89	3·90	3·91	3·92	3·94	3·95	3·96	3·97	3·99
16	4·00	4·01	4·02	4·04	4·05	4·06	4·07	4·09	4·10	4·11
17	4·12	4·14	4·15	4·16	4·17	4·18	4·20	4·21	4·22	4·23
18	4·24	4·25	4·27	4·28	4·29	4·30	4·31	4·32	4·34	4·35
19	4·36	4·37	4·38	4·39	4·40	4·42	4·43	4·44	4·45	4·46
20	4·47	4·48	4·49	4·51	4·52	4·53	4·54	4·55	4·56	4·57
21	4·58	4·59	4·60	4·62	4·63	4·64	4·65	4·66	4·67	4·68
22	4·69	4·70	4·71	4·72	4·73	4·74	4·75	4·76	4·77	4·79
23	4·80	4·81	4·82	4·83	4·84	4·85	4·86	4·87	4·88	4·89
24	4·90	4·91	4·92	4·93	4·94	4·95	4·96	4·97	4·98	4·99
25	5·00	5·01	5·02	5·03	5·04	5·05	5·06	5·07	5·08	5·09
26	5·10	5·11	5·12	5·13	5·14	5·15	5·16	5·17	5·18	5·19
27	5·20	5·21	5·22	5·22	5·23	5·24	5·25	5·26	5·27	5·28
28	5·29	5·30	5·31	5·32	5·33	5·34	5·35	5·36	5·37	5·38
29	5·39	5·39	5·40	5·41	5·42	5·43	5·44	5·45	5·46	5·47
30	5·48	5·49	5·50	5·50	5·51	5·52	5·53	5·54	5·55	5·56
31	5·57	5·58	5·59	5·59	5·60	5·61	5·62	5·63	5·64	5·65
32	5·66	5·67	5·67	5·68	5·69	5·70	5·71	5·72	5·73	5·74
33	5·74	5·75	5·76	5·77	5·78	5·79	5·80	5·81	5·81	5·82
34	5·83	5·84	5·85	5·86	5·87	5·87	5·88	5·89	5·90	5·91
35	5·92	5·92	5·93	5·94	5·95	5·96	5·97	5·97	5·98	5·99
36	6·00	6·01	6·02	6·02	6·03	6·04	6·05	6·06	6·07	6·07
37	6·08	6·09	6·10	6·11	6·12	6·12	6·13	6·14	6·15	6·16
38	6·16	6·17	6·18	6·19	6·20	6·20	6·21	6·22	6·23	6·24
39	6·24	6·25	6·26	6·27	6·28	6·28	6·29	6·30	6·31	6·32
40	6·32	6·33	6·34	6·35	6·36	6·36	6·37	6·38	6·39	6·40
41	6·40	6·41	6·42	6·43	6·43	6·44	6·45	6·46	6·47	6·47
42	6·48	6·49	6·50	6·50	6·51	6·52	6·53	6·53	6·54	6·55
43	6·56	6·57	6·57	6·58	6·59	6·60	6·60	6·61	6·62	6·63
44	6·63	6·64	6·65	6·66	6·66	6·67	6·68	6·69	6·69	6·70
45	6·71	6·72	6·72	6·73	6·74	6·75	6·75	6·76	6·77	6·77
46	6·78	6·79	6·80	6·80	6·81	6·82	6·83	6·83	6·84	6·85
47	6·86	6·86	6·87	6·88	6·88	6·89	6·90	6·91	6·91	6·92
48	6·93	6·94	6·94	6·95	6·96	6·96	6·97	6·98	6·99	6·99
49	7·00	7·01	7·01	7·02	7·03	7·04	7·04	7·05	7·06	7·06
50	7·07	7·08	7·09	7·09	7·10	7·11	7·11	7·12	7·13	7·13
51	7·14	7·15	7·16	7·16	7·17	7·18	7·18	7·19	7·20	7·20
52	7·21	7·22	7·22	7·23	7·24	7·25	7·25	7·26	7·27	7·27
53	7·28	7·29	7·29	7·30	7·31	7·31	7·32	7·33	7·33	7·34
54	7·35	7·36	7·36	7·37	7·38	7·38	7·39	7·40	7·40	7·41

Square roots from 10 to 100

	·0	·1	·2	·3	·4	·5	·6	·7	·8	·9
55	7·42	7·42	7·42	7·44	7·44	7·45	7·46	7·46	7·47	7·48
56	7·48	7·49	7·50	7·50	7·51	7·52	7·52	7·53	7·54	7·54
57	7·55	7·56	7·56	7·57	7·58	7·58	7·59	7·60	7·60	7·61
58	7·62	7·62	7·63	7·64	7·64	7·65	7·66	7·66	7·67	7·67
59	7·68	7·69	7·69	7·70	7·71	7·71	7·72	7·73	7·73	7·74
60	7·75	7·75	7·76	7·77	7·77	7·78	7·78	7·79	7·80	7·80
61	7·81	7·82	7·82	7·83	7·84	7·84	7·85	7·85	7·86	7·87
62	7·87	7·88	7·89	7·89	7·90	7·91	7·91	7·92	7·92	7·93
63	7·94	7·94	7·95	7·96	7·96	7·97	7·97	7·98	7·99	7·99
64	8·00	8·01	8·01	8·02	8·02	8·03	8·04	8·04	8·05	8·06
65	8·06	8·07	8·07	8·08	8·09	8·09	8·10	8·11	8·11	8·12
66	8·12	8·13	8·14	8·14	8·15	8·15	8·16	8·17	8·17	8·18
67	8·19	8·19	8·20	8·20	8·21	8·22	8·22	8·23	8·23	8·24
68	8·25	8·25	8·26	8·26	8·27	8·28	8·28	8·29	8·29	8·30
69	8·31	8·31	8·32	8·32	8·33	8·34	8·34	8·35	8·35	8·36
70	8·37	8·37	8·38	8·38	8·39	8·40	8·40	8·41	8·41	8·42
71	8·43	8·43	8·44	8·44	8·45	8·46	8·46	8·47	8·47	8·48
72	8·49	8·49	8·50	8·50	8·51	8·51	8·52	8·53	8·53	8·54
73	8·54	8·55	8·56	8·56	8·57	8·57	8·58	8·58	8·59	8·60
74	8·60	8·61	8·61	8·62	8·63	8·63	8·64	8·64	8·65	8·65
75	8·66	8·67	8·67	8·68	8·68	8·69	8·69	8·70	8·71	8·71
76	8·72	8·72	8·73	8·73	8·74	8·75	8·75	8·76	8·76	8·77
77	8·77	8·78	8·79	8·79	8·80	8·80	8·81	8·81	8·82	8·83
78	8·83	8·84	8·84	8·85	8·85	8·86	8·87	8·87	8·88	8·88
79	8·89	8·89	8·90	8·91	8·91	8·92	8·92	8·93	8·93	8·94
80	8·94	8·95	8·96	8·96	8·97	8·97	8·98	8·98	8·99	8·99
81	9·00	9·01	9·01	9·02	9·02	9·03	9·03	9·04	9·04	9·05
82	9·06	9·06	9·07	9·07	9·08	9·08	9·09	9·09	9·10	9·10
83	9·11	9·12	9·12	9·13	9·13	9·14	9·14	9·15	9·15	9·16
84	9·17	9·17	9·18	9·18	9·19	9·19	9·20	9·20	9·21	9·21
85	9·22	9·22	9·23	9·24	9·24	9·25	9·25	9·26	9·26	9·27
86	9·27	9·28	9·28	9·29	9·30	9·30	9·31	9·31	9·32	9·32
87	9·33	9·33	9·34	9·34	9·35	9·35	9·36	9·36	9·37	9·38
88	9·38	9·39	9·39	9·40	9·40	9·41	9·41	9·42	9·42	9·43
89	9·43	9·44	9·44	9·45	9·46	9·46	9·47	9·47	9·48	9·48
90	9·49	9·49	9·50	9·50	9·51	9·51	9·52	9·52	9·53	9·53
91	9·54	9·54	9·55	9·56	9·56	9·57	9·57	9·58	9·58	9·59
92	9·59	9·60	9·60	9·61	9·61	9·62	9·62	9·63	9·63	9·64
93	9·64	9·65	9·65	9·66	9·66	9·67	9·67	9·68	9·69	9·69
94	9·70	9·70	9·71	9·71	9·72	9·72	9·73	9·73	9·74	9·74
95	9·75	9·75	9·76	9·76	9·77	9·77	9·78	9·78	9·79	9·79
96	9·80	9·80	9·81	9·81	9·82	9·82	9·83	9·83	9·84	9·84
97	9·85	9·85	9·86	9·86	9·87	9·87	9·88	9·88	9·89	9·89
98	9·90	9·90	9·91	9·91	9·92	9·92	9·93	9·93	9·94	9·94
99	9·95	9·95	9·96	9·96	9·97	9·97	9·98	9·98	9·99	9·99

Answers

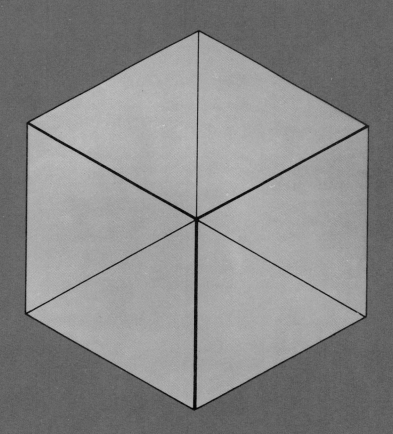

Answers

Algebra—Answers to Chapter 1

Page 3 Exercise 1

1 a, b, d, f are well defined
6 a $A = \{0, 1, 2, \ldots, 10\}, B = \{2, 3, 5, 7\}, C = \{2, 4, 6, 8, 10, 12\}, D = \{3, 6, 9\}$
  b $2 \in A, B, C; 3 \in A, B, D; 5 \in A, B; 10 \in A, C$
     $2 \notin D; 3 \notin C; 5 \notin C, D; 10 \notin B, D$
7 a $\{4, 5, 6\}$ b $\{2, 4, 6\}$ c $\{7, 9, 11\}$ d $\{5, 9, 13, 17, 21\}$
  e $\{(2, 2), (4, 4), (6, 6), (8, 8), (10, 10)\}$ f $\{(6, 3), (10, 5)\}$
8 a T   b F   c T   d F   e T
9 a $S = \{3, 4, 5, \ldots\}$ b $T = \{6, 7, 8, \ldots\}$ c $A = \{-2, -1, 0, 1, 2\}$
  d $B = \{\text{Feb, Apr, Jun, Sep, Nov}\}$
10a $\{x : x > 5, x \in N\}$ b $\{x : x < 10, x \in N\}$ c $\{x : 3 < x < 11, x \in N\}$
  d $\{x : 0 < x < 5, x \in N\}$ e $\{x : x \text{ is even}, x \in N\}$ f $\{x : x \text{ is prime}, x \in N\}$

Page 5 Exercise 2

1 a $F = \{1, 3, 5, 7, 9\}, G = \{11, 13, 17, 19, 23\}, H = \{7, 14, 21, 28, 35, 42, 49\}$
  b $n(F) = 5, n(G) = 5, n(H) = 7; F \text{ and } G$; no
2 Every member of B is not a member of A. $n(A) = 4$ and $n(B) = 5$.
3 $n(C) = 10$ 4a no   b yes
5 a 4, 3, 1, 1, 0, 3   b $B = Q$
6 $S = \{541, 514, 451, 415, 154, 145\}; n(S) = 6$
7 4, 8, 2, 6, 2, 12
8 $P = \{2, 3, 5, 7, 11\}; Q = \{2, 3, 5, 7, 11\}$. Hence $P = Q$.
9 a, b, d are finite; c, e, f are infinite. Note: ø is finite.
10a $\{3, 6, 9, \ldots, 99\}$ b $\{1, 3, 5, \ldots\}$ c $\{-24, -23, -22, \ldots, -1\}$
  d $\{\frac{1}{1}, \frac{1}{2}, \frac{1}{3}, \ldots\}$.

Page 7 Exercise 3

1 $C = D, C \subset D, D \subset C, F \subset A, F \subset B, F \subset C, F \subset D, F \subset G,$
   $B \subset A, C \subset A, D \subset A, C \subset G, D \subset G$
2 a T   b F   c T   d T   e T   f F
3 a {Jan, Mar, May, Jul, Aug, Oct, Dec}
  b {Apr, Jun, Sep, Nov}   c {Feb}
5 a $S \subset R$ b $P \subset Z$ c $R \subset P$
6 a $P = \{2, 3, 5, 7\}$ b $Q = \{1, 3, 5, 7, 9\}$. no   c $S = \{1, 2, 3, \ldots, 10\}$
  d { }
7 a (1) $\{a\}$, ø   (2) $\{a, b\}, \{a\}, \{b\}$, ø
     (3) $\{a, b, c\}, \{a, b\}, \{b, c\}, \{c, a\}, \{a\}, \{b\}, \{c\}$, ø
     (4) $\{a, b, c, d\}, \{a, b, c\}, \{a, b, d\}, \{a, c, d\}, \{b, c, d\}, \{a, b\}, \{a, c\}, \{a, d\},$
         $\{b, c\}, \{b, d\}, \{c, d\}, \{a\}, \{b\}, \{c\}, \{d\}$, ø
  b (1) $2^5 = 32$ (2) $2^6 = 64$ c 2^n d 7
8 a $P = Q$ b ø   c $A \subset C$

   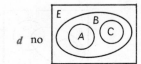 d no

Answers

Page 9   Exercise 4

1 a $A \cap B = \{4, 5\} = B \cap A$ **b** sets equal; commutative law

c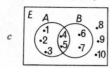

2 a $P \cap Q = \{1, 2, 3\} = Q \cap P$ **b** P **c**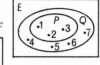

3 a $C \cap D = \{\ \}$ or ø   **b**

4 $A = \{1, 3, 5, 7, 9\}, A \cap B = \{7, 9\}, A \cap C = \{1, 3\}, B \cap C = $ ø
5 $P \cap Q = \{b, c, d\}, Q \cap R = \{d\}, R \cap S = \{f\}, P \cap R = \{d, e, f\},$
  $Q \cap S = \{\ \}, P \cap P = \{a, b, c, d, e, f\}, Q \cap Q = \{b, c, d\}, S \cap S = \{f\}$
6 a $A \cap B = \{3, 4\}, B \cap C = \{5\}, A \cap C = $ ø$, E \cap B = \{3, 4, 5\}$

b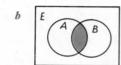

  $A \cap B \neq $ ø $B \cap C = C$ $A \cap C = $ ø $E \cap B = B$

7   Some right-angled triangles are isosceles.

8 $X = \{0, 6, 12, 18, 24, 30\},$
  $Y = \{0, 8, 16, 24, 32\},$
  $X \cap Y = \{0, 24\}$ which is the set of common multiples. 24.

9 a $X = \{0, 4, 8, 12, 16, 20, 24, 28\}, Y = \{0, 6, 12, 18, 24, 30\},$
  $X \cap Y = \{0, 12, 24\}$. LCM of 4 and 6 = 12.
b $X = \{0, 9, 18, 27, 36, 45\}, Y = \{0, 12, 24, 36, 48\}, X \cap Y = \{0, 36\}$. LCM of 9 and 12 = 36.
10 The set of yellow roses in the flower shop. If the shop has no yellow roses.

Page 10   Exercise 4B

1 a (1) $A \cap B = \{2, 4, 5\}$ (2) $B \cap C = \{2, 4, 6\}$ (3) $\{2, 4\}$ (4) $\{2, 4\}$
  b $(A \cap B) \cap C = A \cap (B \cap C)$
2 c yes, the associative law   **3** three
4 $x = 9$ **5** $x = 11$ **6b** 8   **c** 40   **d** 30

7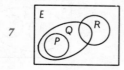

Answers

Page 13 Exercise 5

1 a (i) {1, 2, 3, 4, 5, 6} (ii) {1, 2, 3, 4} (iii) {p, q, r, s}   **b** yes
c (i) {3} (ii) {1, 2} (iii) ø

2 (i)　　　　　　(ii)　　　　　　(iii)　　　　　　(iv) $P \cap Q = \emptyset$

d yes, commutative law

3 a $P \cup Q = \{1, 2, 3, 5, 7\}$ **b** $X \cup Y = \{0, 1, 2, 3, 4, 5\}$
c $S \cup T = \{a, b, c\}$ **d** $A \cup B = \{a, b, c, d, e\}$

4 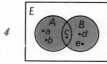 $n(A \cup B) = 5$

5 $P \cup Q = \{1, 2, 3, 4, 5, 6, 7, 8,\}\ n(P \cup Q) = 8$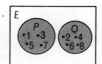

6 the set of pupils in the school

7 a (*1*) {letters in the alphabet} (*2*) empty set   **b**(*1*) 26 (*2*) 0

8 $X = \{2, 3, 5\},\ Y = \{2, 3, 5, 7\}$. No. $X \cup Y = \{2, 3, 5, 7\},\ X \cap Y = \{2, 3, 5\}$

9 a R   **b** S   **10a** 86   **b** 18   **c** 25

Page 14 Exercise 5B

1 a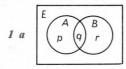

b (*1*) $p+q+r$ (*2*) $p+q+r$ **c** $n(A \cup B) = n(A)+n(B)$
2 (*11*) $A \cup B$ (*12*) $B \cup C$ (*13*) $A \cup B \cup C$ (*14*) $B \cap C$ (*15*) $(A \cap B) \cup C$
    (*16*) $C \cap (A \cup B)$ or $(C \cap A) \cup (C \cap B)$

3 a {0, 1, 2, 3, 5, 7}   **b** {0, 1, 2, 3, 5, 8}   **c** {1, 2, 3, 5, 7, 8}
  d {0, 1, 2, 3}   **e** {0, 1, 2, 3}   **f** {0, 1, 2, 3}
4 a {0, 1, 2, 3, 5, 7, 8}   **b** {0, 1, 2, 3, 5, 7, 8}. They are the same.
5 c yes; the associative law for the union of sets
6 a (*1*) {1, 2, 3, 4, 5, 6, 8, 10}   (*2*) {2, 3, 4, 6, 8, 9, 10, 12}
    (*3*) {1, 2, 3, 4, 5, 6, 9, 12}   (*4*) {2, 4, 6}
    (*5*) {6}   (*6*) {3, 6}
  b (*1*) {2, 3, 4, 6}   (*2*) {1, 2, 3, 4, 5, 6}
    (*3*) {2, 3, 4, 6}   (*4*) {1, 2, 3, 4, 5, 6}
  c (*1*) and (*3*); (*2*) and (*4*)
7 a $A \cap (B \cup C) = (A \cap B) \cup (A \cap C)$ **b** $A \cup (B \cap C) = (A \cup B) \cap (A \cup C)$
8 40

Answers

Page 16 Exercise 6

1. $X' = \{q, s, t\}$
2. (1) $P = \{2, 3, 5, 7, 11, 13, 17, 19\}$
   (2) $P' = \{1, 4, 6, 8, 9, 10, 12, 14, 15, 16, 18, 20\}$
3. $T' = \{\text{odd numbers}\}$
4. {all non-negative integers}, or {positive integers and zero}
5. A' is the set of all triangles which are not isosceles
6. $A' = \{x : x < 2\}$, $B' = \{x : x \geq 5\}$
7. $A = \{\text{rectangles, squares, rhombuses, kites}\}$, $A' = \{\text{parallelograms}\}$,
   $B = \{\text{rectangles, squares, parallelograms, rhombuses}\}$, $B' = \{\text{kites}\}$
8. $A' = \{4, 6, 8\}$, $B' = \{2, 4, 6, 8\}$, $C' = \{1, 3, 5, 7\}$.
9.
   a T   b F   c F   d T   e T   f F   g F   h T
10. a $\{1, 3, 5, 7\}$ b $\{2, 4, 6, 8\}$ c $\{2, 5, 7\}$
    d $\{1, 3, 4, 6, 8\}$ e $\{5, 7\}$ f $\{1, 2, 3, 4, 6, 8\}$
    g $\{1, 2, 3, 5, 7\}$ h $\{4, 6, 8\}$ i $\{1, 2, 3, 4, 6, 8\}$
    j $\{4, 6, 8\}$ f and i; h and j
11. (i) A' (ii) B' (iii) $A \cap B$ (iv) $A \cup B$

Page 18 Exercise 6B

1. a $\{p, q, r, s\}$ b $\{x, y, z, t\}$ c $\{r, s, t, x\}$ d $\{p, q, y, z\}$
   e $\{p, q\}$ f $\{x, t\}$. $\{p, q, x, t\}$
2. $P \cap Q'$, $P \cap Q$, $Q \cap P'$, $(P \cup Q)'$ or $P' \cap Q'$ 3. P, $Q \cap P'$, Q'
4. I—{quadrilaterals which have bilateral but not half turn symmetry}
   II—{quadrilaterals which have both bilateral and half turn symmetry}
   III—{quadrilaterals which have half turn but not bilateral symmetry}
   IV—{quadrilaterals which have neither bilateral nor half turn symmetry}
   I—kite; II—rectangle, or square, or rhombus; III—parallelogram;
   IV—quadrilaterals with unequal sides
5. a $\{x : x \notin A\}$ b $\{x : x \in A \text{ and } x \in B\}$ c $\{x : x \in A \text{ and } x \notin B\}$
6. $(P \cup Q)' = P' \cap Q'$ 7. $(P \cap Q)' = P' \cup Q'$

Algebra—Answers to Chapter 2

Page 24 Exercise 1

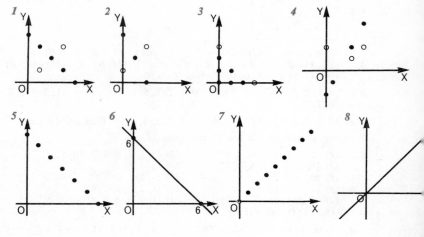

Answers

9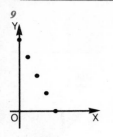
10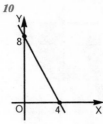
11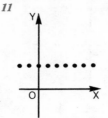
12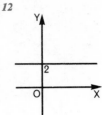

13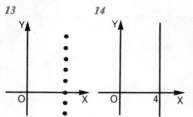
14
15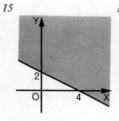
16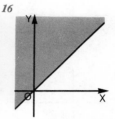

Page 26 Exercise 2

1. $2x+y-4 = 0$; $2x+y = 4$; $y = -2x+4$
2. $2x-y-1 = 0$; $2x-y = 1$; $y = 2x-1$
3. $x+3y-6 = 0$; $x+3y = 6$; $y = -\frac{x}{3}+2$
4. $x+2y-3 = 0$; $x+2y = 3$; $y = -\frac{x}{2}+\frac{3}{2}$
5. $x-y-5 = 0$; $x-y = 5$; $y = x-5$
6. $2x-y+1 = 0$; $2x-y = -1$; $y = 2x+1$
7. $2x-3y-6 = 0$; $2x-3y = 6$; $y = \frac{2}{3}x-2$
8. $3x+2y+6 = 0$; $3x+2y = -6$; $y = -\frac{3}{2}x-3$
9. $3x-4y-12 = 0$; $3x-4y = 12$; $y = \frac{3}{4}x-3$
10. $x-2y+2 = 0$; $x-2y = -2$; $y = \frac{x}{2}+1$
11. $x+0y+2 = 0$; $x+0y = -2$; $0y = x+2$
12. $0x+y-3 = 0$; $0x+y = 3$; $y = 0x+3$

13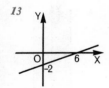
14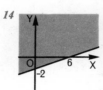
15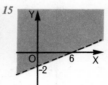
16

17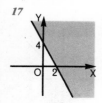
18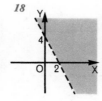
19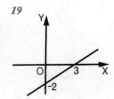
20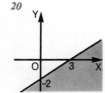

Answers

21–30 (graphs)

Page 27 Exercise 3

1–6 (graphs)

7 The solution sets do not intersect.

8–12 (graphs)

6: $x+2y>8$

9: 4 inequations

Answers

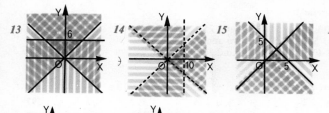
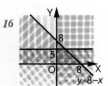
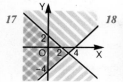

Page 29   Exercise 4

1	{(3, 4)}	2	{(0, −2)}	3	{(4, 3)}	4	{(−3, 9)}
5	{(4, 4)}	6	{(1, 3)}	7	ø	8	{(1, −1)}
9	{(0·5, 0·5)}	10	{(1·3, 0·7)}	11	{(1·6, 0·6)}	12	{(−1·3, 0·7)}
13	{(2·3, 1·3)}	14	{(−1·5, 1·5)}	15	{(−1·3, 2·7)}	16	{ } or ø
17	{(2·5, 0·5)}	18	{(1, 0·6)}	19	{(3, 2)}	20	{(3, −1)}

Page 32   Exercise 5

1	{(8, 4)}	2	{(3, 3)}	3	{(3½, 1½)}	4	{(1, 1)}
5	{(2, 0)}	6	{(−1, 1)}	7	{(1, 1)}	8	{(3, 2)}
9	{(2, −1)}	10	{(−1, −1)}	11	{(3, −3)}	12	{(3, 1)}
13	{(3/2, 1)}	14	{(3, −2)}	15	{(4, 6)}	16	{(7, −2)}
17	{(−1, 2)}	18	{(2, 1)}	19	{(−8, −5)}	20	{(4/3, 2/3)}
21	{(8/5, 3/5)}	22	{(−4/3, 2/3)}	23	{(−8, −9)}	24	{(35/37, 28/37)}
25	{(3/2, 1/2)}	26	{(−2, 5)}	27	{(3, −2)}	28	{(0·4, 0·7)}

Page 33   Exercise 6

1	{(5, 5)}	2	{(2, 4)}	3	{(12, 4)}	4	{(−6, −3)}
5	{(2, 1)}	6	{(7, 3)}	7	{(−1, 2)}	8	{(10, 5)}
9	{(2, 3)}	10	{(5, −2)}	11	{(−1, −2)}	12	{(3, 4)}
13	{(−2, −2)}	14	{(3/2, −3)}	15	{(−1/3, 1/3)}	16	{(3, 2)}
17	{(½, 2)}	18	{(1, −2)}	19	{(12, −50)}	20	{(4, 3)}

Page 34   Exercise 7

1	{(4, 1)}	2	{(5, 0)}	3	{(3, −1)}	4	{(4, 2)}
5	{(6, 6)}	6	{(2, −3)}	7	{(3, 2)}	8	{(3, 8)}
9	{(−3, 6)}	10	{(6, −10)}				

Page 34   Exercise 7B

1	{(4, 4)}	2	{(6, −3)}	3	{(−3, 2)}	4	ø
5	{(4, 7)}	6	{(−1, 8)}	7	{(4, 3)}	8	{(7, 5)}
9	{(10½, −3)}	10	{(4, 7)}	11	{(2, 3)}	12	{(−4, 1)}

Answers

Page 36 Exercise 8

1. 74 and 38
2. $37\frac{1}{2}$ and $25\frac{1}{2}$
3. $73\frac{1}{2}°$ and $16\frac{1}{2}°$
4. 124° and 56°
5. 51 cm and 33 cm
6. $30\frac{1}{2}$ cm and $3\frac{1}{2}$ cm
7. 30p and 40p
8. 25p and 10p
9. $m = 4, c = -6; a = 5$
10. $m = 3, c = 1$. no

Page 37 Exercise 8B

1. $a = 3, b = 4$
2. $p = 3, q = 4$
3. $d = 6, k = -3$; 597
4. $a = -\frac{3}{4}, b = 16$ a $V = 8\frac{1}{2}$ b $t = 21\frac{1}{3}$
5. $a = 24, b = -5$; $h = 16$; projectile lands on the ground
6. $a = 2, b = -3$
7. 140 and 84
8. $x = 68t; x = 72(t - \frac{1}{6})$ a 3 hours   b 204 km
9. $x = \frac{1}{3}(p+2q)$ 10 $x = a+b, y = b-a$
11. 175 front stalls, 305 back stalls   12 12 and 6 hours respectively

Algebra—Answers to Chapter 3

Page 42 Exercise 1

1. a $P = 2x+2y$ or $P = 2(x+y)$ and $A = xy$ b 40 cm and 96 cm²
2. a $P = 4s$ and $A = s^2$ b 60 cm and 225 cm²
3. a $d = x+y$ b $d = 1300$ 4a $d = 3(x+y)$ b $d = 1515$
5. $W = x+y$ a $W = 5.25$ b $y = 2\frac{1}{2}$
6. $L = p-q; L = 3.5$
7. $P = 2x+2y+z$ a $P = 112$ b $z = 20$ c $x = 7$
8. a $S = 4a+40$ or $S = 4(a+10)$ b $S = 100$
9. a $\{a\}, \emptyset; 2$ b $\{a\}, \{b\}, \{a, b\}, \emptyset; 4 = 2^2$
   c $\{a\}, \{b\}, \{c\}, \{a, b\}, \{b, c\}, \{c, a\}, \{a, b, c\}, \emptyset; 8 = 2^3$ d $16 = 2^4$ e 2^n
10. a second row: 1 2 3 4 5 7 17 $n-3$ b $d = n-3$ c 97

Page 44 Exercise 1B

1. a $b = 2x$ b $d = 3x+2y$ c $e = 4x+2y$ d $f = 4x+3y$
   150 m, 325 m, 400 m, 450 m
2. a $P = 6x+\pi x$ or $P = x(6+\pi)$
   b $A = 4x^2+\frac{1}{2}\pi x^2$ or $A = x^2(4+\frac{1}{2}\pi); P = 91.4$ and $A = 557$
3. a 19, 23, 27   c 399 and 3999
4. a 12   b 20   c (1) $S = xy$ (2) $P = (x+1)(y+1)$ (3) 870; 930
5. a 14   c 338350   6a 2, 5   b 9   d 170   e 12

Page 47 Exercise 2

1. $\{\frac{5}{2}\}$
2. $\{\frac{8}{3}\}$
3. $\{-\frac{1}{2}\}$
4. $\{-\frac{4}{5}\}$
5. $\{7\}$
6. $\{1\}$
7. $\{-3\}$
8. $\{-11\}$
9. $\{3\}$
10. $\{\frac{5}{2}\}$
11. $\{\frac{5}{4}\}$
12. $\{\frac{1}{2}\}$
13. $\{8\}$
14. $\{-3\}$
15. $\{\frac{15}{2}\}$
16. $\{-12\}$
17. $\{\frac{1}{2}\}$
18. $\{\frac{1}{3}\}$
19. $\{2\}$
20. $\{\frac{1}{2}\}$
21. $\{-\frac{1}{2}\}$
22. $\{0\}$
23. $\{6\}$
24. $\{4\}$
25. $\{-10\}$
26. $\{-1\}$
27. $\{1\}$
28. $\{9\}$
29. $\{2\}$
30. $\{-30\}$
31. $\{-\frac{1}{2}\}$
32. $\{2\}$
33. $\{\frac{5}{2}\}$
34. $\{-1\}$
35. $\{1\}$

Page 48 Exercise 3

1. $\{3\}$
2. $\{7-a\}$
3. $\{b-3\}$
4. $\{b-a\}$
5. $\{11\}$
6. $\{a+5\}$
7. $\{4+c\}$
8. $\{p+q\}$
9. $\{3h\}$
10. $\{6c\}$
11. $\{a\}$
12. $\{5a\}$

Answers

13 $\left\{\dfrac{5}{2}\right\}$ 14 $\left\{\dfrac{a}{3}\right\}$ 15 $\left\{\dfrac{4}{a}\right\}$ 16 $\left\{\dfrac{d}{c}\right\}$

17 $\left\{-\dfrac{q}{p}\right\}$ 18 $\left\{-\dfrac{f}{d}\right\}$ 19 $\left\{-\dfrac{h}{k}\right\}$ 20 $\left\{\dfrac{1}{2}\right\}$

21 $\left\{\dfrac{a}{2}\right\}$ 22 $\left\{\dfrac{4}{a}\right\}$ 23 $\left\{\dfrac{b-a}{2}\right\}$ 24 $\left\{\dfrac{h-r}{4}\right\}$

25 $\left\{\dfrac{5}{3}\right\}$ 26 $\left\{\dfrac{5a}{3}\right\}$ 27 $\left\{\dfrac{c+d}{2}\right\}$ 28 $\left\{\dfrac{4}{a}\right\}$

29 $\left\{\dfrac{b+7}{a}\right\}$ 30 $\left\{\dfrac{q+r}{p}\right\}$ 31 $\left\{\dfrac{1}{2}\right\}$ 32 $\left\{\dfrac{5b}{3}\right\}$

33 $\left\{\dfrac{d-4c}{4}\right\}$ 34 $\left\{\dfrac{5h+k}{5}\right\}$ 35 $\left\{\dfrac{c-ab}{a}\right\}$ 36 $\left\{\dfrac{pq+r}{p}\right\}$

37 {6} 38 {5a} 39 {4c} 40 {ab}

41 $\left\{\dfrac{3}{2}\right\}$ 42 $\left\{\dfrac{ab}{2}\right\}$ 43 {6} 44 $\left\{\dfrac{4h}{3}\right\}$

45 $\left\{\dfrac{5b}{a}\right\}$ 46 $\left\{\dfrac{pq}{c}\right\}$ 47 $\left\{\dfrac{ab}{c}\right\}$ 48 $\left\{\dfrac{r}{k}\right\}$

49 $\{10-a\}$ 50 $\{bc-a\}$ 51 $\{p+3q\}$ 52 $\{p+qr\}$

Page 49 Exercise 3B

1 $\{4-a\}$ 2 $\{a-3\}$ 3 $\left\{\dfrac{1-a}{2}\right\}$ 4 $\left\{\dfrac{a-5}{4}\right\}$

5 $\left\{\dfrac{a+b}{2}\right\}$ 6 $\left\{\dfrac{b-a}{2}\right\}$ 7 $\left\{\dfrac{a+b}{2}\right\}$ 8 $\left\{\dfrac{3}{a+b}\right\}$

9 $\left\{\dfrac{a-c}{b}\right\}$ 10 $\left\{\dfrac{r-pq}{p}\right\}$ 11 $\left\{\dfrac{ac+b}{a}\right\}$ 12 $\left\{\dfrac{y-c}{m}\right\}$

13 $\{y-3\}$ 14 $\left\{\dfrac{3-y}{2}\right\}$ 15 $\left\{\dfrac{5y+10}{2}\right\}$ 16 $\{-4y-5\}$

17 $\left\{\dfrac{3y+4}{2}\right\}$ 18 $\{a-y\}$ 19 $\left\{\dfrac{c-by}{a}\right\}$ 20 $\left\{\dfrac{a(b-y)}{b}\right\}$

21 $\{-a\}$ 22 $\{b-2a\}$ 23 $\left\{\dfrac{5}{a+b}\right\}$ 24 $\{0\}$

25 $\left\{\dfrac{d-b}{a-c}\right\}$ 26 $\left\{\dfrac{ab}{a-c}\right\}$ 27 $\left\{\dfrac{ah+bk}{a-b}\right\}$ 28 $\left\{\dfrac{a}{1-n}\right\}$

29 $\left\{\dfrac{p}{q-1}\right\}$ 30 $\left\{\dfrac{b}{a-b}\right\}$

Page 51 Exercise 4

1 {12} 2 {2} 3 {2} 4 {6} 5 $\{\tfrac{9}{2}\}$
6 $\{\tfrac{5}{2}\}$ 7 {18} 8 $\{\tfrac{15}{4}\}$ 9 {8} 10 $\{3\}$
11 $\{\tfrac{15}{4}\}$ 12 {9} 13 {16} 14 {4} 15 $\{\tfrac{3}{2}\}$
16 $\{-\tfrac{3}{2}\}$ 17 $\{-\tfrac{1}{2}\}$ 18 $\{\tfrac{1}{10}\}$ 19 $\{\tfrac{11}{10}\}$ 20 $\{\tfrac{3}{2}\}$

Page 51 Exercise 4B

1 {7·6} 2 {1·44} 3 {10} 4 $\left\{\dfrac{15}{2}\right\}$

5 $\left\{\dfrac{a}{2}\right\}$ 6 $\left\{\dfrac{3b}{a}\right\}$ 7 $\{a^2\}$ 8 $\left\{\dfrac{bd}{c}\right\}$

Answers

9 $\left\{\dfrac{1}{a+1}\right\}$ 10 $\left\{\dfrac{a-b}{c}\right\}$ 11 $\left\{\dfrac{a}{5}\right\}$ 12 $\left\{\dfrac{3q}{p}\right\}$

13 $\{7a\}$ 14 $\left\{\dfrac{5b}{a}\right\}$ 15 $\left\{\dfrac{pq}{1-q}\right\}$ 16 $\left\{\dfrac{ma}{m+n}\right\}$

Page 53 Exercise 5

1   $x = \dfrac{P}{4}$ 2 $l = \dfrac{A}{b}$ 3 $h = \dfrac{V}{lb}$

4 a $D = ST$ b $T = \dfrac{D}{S}$ 5a $V = IR$ b $R = \dfrac{V}{I}$

6   $h = \dfrac{2A}{b}$ 7 $x = \sqrt{A}$ 8 $r = \sqrt{\dfrac{A}{3 \cdot 14}}$ 9 $a = \dfrac{2A}{3b}$

10   $m = \dfrac{y-c}{x}$ 11a $C = Id^2$ b $d = \sqrt{\dfrac{C}{I}}$ 12 $r = \dfrac{D}{2(\pi+1)}$

13a $d = \dfrac{n-2a}{3}$ b   6   14a $b = \dfrac{P-2l}{2}$ b   4

15a $c = \sqrt{(a^2-b^2)}$ b   6   16a $l = \dfrac{2S-na}{n}$ b   2·5

17a $n = \dfrac{R+4}{2}$; 7   b   no, since n must be a whole number

18a $h = \dfrac{V}{\pi r^2}$ b $r = \sqrt{\dfrac{V}{\pi h}}$; 0·87 approx.

Page 54 Exercise 5B

1 a $f = \dfrac{v-u}{t}$ b $u = \dfrac{s - \frac{1}{2}ft^2}{t}$ or $u = \dfrac{2s - ft^2}{2t}$ c $t = \dfrac{2s}{u+v}$

d $s = \dfrac{v^2 - u^2}{2f}$ e $f = \dfrac{s-ut}{\frac{1}{2}t^2}$ or $f = \dfrac{2s-2ut}{t^2}$ f $u = \sqrt{(v^2 - 2fs)}$

2   $m = \sqrt{ab}$ 3 $h = \dfrac{3L+c}{8}$ 4a $d = \dfrac{T-a}{n-1}$ b $n = \dfrac{T+d-a}{d}$

5   $r = \sqrt[3]{\dfrac{3V}{4\pi}}$ 6 $h = \dfrac{A - 2\pi r^2}{2\pi r}$ 7 $x = \dfrac{y-1}{y+1}$

8 a $s = \dfrac{c(p+100)}{100}$ b $c = \dfrac{100s}{p+100}$ 9b   1302   c $l = \dfrac{2S-na}{n}$; 17

10a   64   b   60; 62; 63; 63·5; 63·75; 63·875; 63·938   c $r = \dfrac{S-a}{S}$; $\frac{1}{3}$.

Algebra—Answers to Revision Exercises

Page 57 Revision Exercise 1A

1 a   yes, finite   b   yes, infinite   c   no   d   yes, finite or empty   e   no

2 a $\{1, 2, 3, \ldots, 50\}$ b $\{a, e, i, o, u\}$ c $\{-24, -23, -22, \ldots, -1\}$

d $\{3, 6, 9, \ldots\}$

3 a $\{E, H, I, O, X\}$ b $\{A, H, I, M, O, T, V, W, X, Y\}$ c $\{H, I, O, X\}$

4 a $B \subset A$, $B = D$ are true, the others are false.

b   (1) 45°, 45°, 90°   (2) sides equal

5 a $\{2, 4, 6, 8, \ldots, 20\}$ b $\{4, 8, 12, 16, 20\}$ c $\{2, 3, 5, 7, 11, 13, 17, 19\}$

Answers

6 *a* For example, {a, b, d}, {b, c, d}, {c, a, e}
b For example, {a, b, c, d}, {b, c, d, e}, {a, c, d, e}, {b, c, e, a}   *c* For example, ø

7 *a* $H \cap K = \{4\}$, $H \cup K = \{1, 2, 3, 4, 5, 6, 7, 8\}$ *b*
c $n(H \cap K) + n(H \cup K) = 1 + 8 = 9$
$n(H) + n(K) = 4 + 5 = 9$ so L.S. = R.S.

8 *a* {3}   *b* ø   *c* {1, 2, 3, 4}   *d* {1, 2, 3}
9 *a* {p, q, r}   *b* {r, s}   *c* {x, y, s}   *d* {p, q, x, y}
e {r}   *f* {p, q, r, s}   *g* {x, y}   *h* {p, q, s, x, y}
i {x, y}   *j* {p, q, s, x, y}
$(A \cap B)' = A' \cup B'$ and $(A \cup B)' = A' \cap B'$.
10*a* {a, b}   *b* {a, b, c, d, e}   *c* {c, d, e, f, g, h}   *d* {f, g, h}
e {a, b}   *f* {a, b, c, d, e}   *g* {f, g, h}   *h* {c, d, e, f, g, h}
i {f, g, h}   *j* {c, d, e, f, g, h}
$A = A \cap B$, $A \cup B = B$, $A' = A' \cup B' = (A \cap B)'$,
$B' = A' \cap B' = (A \cup B)'$

Page 59 Revision Exercise 1B

1 $\bar{0} = \{4, 8, 12, \ldots, 32\}$, $\bar{1} = \{1, 5, 9, \ldots, 29\}$
$\bar{2} = \{2, 6, 10, \ldots, 30\}$, $\bar{3} = \{3, 7, 11, \ldots, 31\}$
a $\bar{3}$ *b* $\bar{3}$ *c* $\bar{1}$ *d* $\bar{0}$
2 $P = T$, $P \subset S$, $T \subset S$, $Q = R$, $Q \subset V$, $R \subset V$
3 *a* {5, 6, 7}   *b* {1, 2, 3, ..., 10}   *c* {8, 9, 10, 11, 12}
d {1, 2, 3, 4, 11, 12}   *e* {1, 2, 3, 4, 8, 9, 10, 11, 12}   *f* {11, 12}
g {1, 2, 3, 4}   *h* {8, 9, 10}

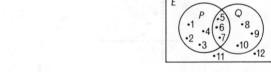

4

∪	ø	A	B	E
ø	ø	A	B	E
A	A	A	E	E
B	B	E	B	E
E	E	E	E	E

∩	ø	A	B	E
ø	ø	ø	ø	ø
A	ø	A	ø	A
B	ø	ø	B	B
E	ø	A	B	E

Symmetry about the main diagonal.

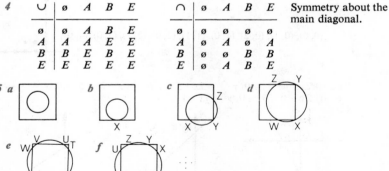

Note. The elements of the sets are given by the letters in each diagram,

Answers

6 a $(A \cap C) \cup B$ or $(A \cap B) \cup (A \cap C)$ b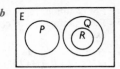

7 a (1) $P = \{0, 1, 2, 3, 4\}$ (2) $Q = \{-3, -2, -1, 0, 1, 2\}$
     (3) $R = \{-2, -1, 0, 1, 2, 3\}$
  b $a = -3$ and $b = 4$
  c $Q \cup R = \{-3, -2, -1, 0, 1, 2, 3\}$, so $P \cap (Q \cup R) = \{0, 1, 2, 3\}$;
     $P \cap Q = \{0, 1, 2\}$ and $P \cap R = \{0, 1, 2, 3\}$, so
     $(P \cap Q) \cup (P \cap R) = \{0, 1, 2, 3\}$.

8 a 90   b 100   c 69   d 16   9 95   10 18

11a ... b ... c AuD d AnB
  e A'nC' f (AuD)' g ∅ h (AnB)uD

12a VIII   b VII   c V   d III   e VI   f IV   g II   h I

Page 62   Revision Exercise 2A

1 $\{(0, 7), (1, 5), (2, 3), (3, 1)\}$

2 $\{(0, 0), (0, 1), (0, 2), (1, 0), (1, 1), (2, 0), (2, 1), (3, 0), (4, 0)\}$

3 a (i) $2x + y - 3 = 0$, $y = -2x + 3$ b (i) $x + 2y + 5 = 0$, $y = -\frac{1}{2}x - \frac{5}{2}$
  c (i) $2x - 3y - 9 = 0$, $y = \frac{2}{3}x - 3$

4 a   b   c

Answers

5 **6a** 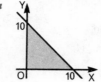 **b**

c **d**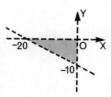

7 From the graphs the solution sets are: *a* $\{(7, -2\cdot7)\}$ *b* $\{(5, 0)\}$
8 *a* $\{(2, 0)\}$ *b* $\{(-2, 1)\}$ *c* $\{(-1, 1)\}$ *d* $\{(-3, 4)\}$
9 *a* $\{(4, 2)\}$ *b* $\{(1\frac{1}{2}, 1)\}$
10a A(4, 0), B(0, 8)   *b* I, II and VI   *c* $\{(x, y): x \geq 0, y \geq 0, 2x+y \leq 8, x, y \in R\}$
11 (i) $y < 3$ (ii) $3x - 2y \leq 0$, or $y \geq \frac{3}{2}x$ **12** $a = -1, b = -2$
13 $p = 6, q = -9$ **14** $\{(1\frac{1}{2}, -1)\}$

Page 63 Revision Exercise 2B

1 a ... **c/d**

2 a/b   **c/d**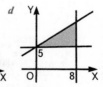

3 a (i) $5x - 2y - 6 = 0$ (ii) $y = \frac{5}{2}x - 3$
  b (i) $3x + 7y - 5 = 0$ (ii) $y = -\frac{3}{7}x + \frac{5}{7}$
  c (i) $6x + 4y - 3 = 0$ (ii) $y = -\frac{3}{2}x + \frac{3}{4}$
4 From the graphs the solution sets are: *a* $\{(1, 3)\}$ *b* ø   *c* $\{(0, 2)\}$
5 *a* $\{(-6, 6)\}$ *b* $\{(2, -5)\}$ *c* $\{(1, 3)\}$
6 *a* $\{(2, 3)\}$ *b* $\{(5, 7)\}$ *c* $\{(4, -2)\}$
7 *a* P(0, 3), Q(5, 3), R(5, 8)   *b* $\{(x, y): x \leq 5, y \geq 3, y \leq x+3, x, y \in R\}$
  c $\{(x, y): y \geq x+3, x, y \in R\}$
8 $x + 3y \leq 6$, or $y \leq -\frac{1}{3}x + 2$ **9** $m = 3, c = -1$
10 $2\frac{1}{2}$ **11** $p = \frac{5}{4}, q = -10$ *a* 65   *b* 40
12 $26\frac{1}{2}$ **13** 48   **14** $a = 2, b = 6$; 20600

Answers

Page 65 Revision Exercise 3A

1 a $L = 4x+8y$; 74   b $A = 4xy+2y^2$; 220   c $V = xy^2$; 212·5
2 $A = ab - 2c^2$; 750.
3 a $\{q-p\}$ b $\{4p\}$ c $\left\{\dfrac{1}{p}\right\}$ d $\{2n\}$
  e $\{ct\}$ f $\left\{\dfrac{a-b}{2}\right\}$ g $\{a-b\}$ h $\left\{\dfrac{pq+r}{p}\right\}$
4 a 4, 9, 16, 25   b $N = n^2$; 100
5 a $y = 2x-3$ b $y = \dfrac{3-4x}{2}$ c $y = \dfrac{3x-6}{2}$ d $y = \dfrac{8-x}{2}$
  e $y = \tfrac{5}{4}x$ f $y = \dfrac{6-2x}{3}$
6 $m = \tfrac{4}{5}n$, $m = 960$
7 a $r = \dfrac{C}{2\pi}$ b $r = \sqrt{\dfrac{A}{\pi}}$ c $r = \sqrt{\dfrac{V}{\pi h}}$ d $r = \dfrac{v^2}{a}$ e $r = \sqrt[3]{\dfrac{3V}{4\pi}}$
8 $P = 9a+13b$; 41·0
9 a $m = \dfrac{Ft}{v-u}$ b $v = \dfrac{Ft+mu}{m}$ c $u = \dfrac{mv-Ft}{m}$ 10 $d = \sqrt{\dfrac{kL}{R}}$; 3

Page 66 Revision Exercise 3B

1 a $\left\{\dfrac{r-p}{q}\right\}$ b $\left\{\dfrac{a+b}{2}\right\}$ c $\left\{\dfrac{p}{m+n}\right\}$ d $\left\{\dfrac{c-b}{a-1}\right\}$ e $\left\{\dfrac{abc}{c-a}\right\}$
  f $\{-ab\}$ g $\left\{\dfrac{2ab}{a+b}\right\}$ h $\{0\}$ i $\left\{\dfrac{2mn}{m-n}\right\}$

2 a n; 10   b 3^n; 59049   c $\dfrac{1}{n(n+1)}$; $\dfrac{1}{110}$

3 $p = m - \dfrac{cm}{100} = m\left(1 - \dfrac{c}{100}\right)$ 4 a $(L-x)$ metres   b $x = \dfrac{mL}{m+n}$

5 a $y = 3-2x$ b $y = 2x-5$ c $y = \dfrac{x+9}{3}$
  d $y = -\dfrac{2x+12}{3}$ e $y = \dfrac{12-4x}{3}$ f $y = \dfrac{4x-15}{3}$

6 a $y = 180-2x$ b $x = \dfrac{180-y}{2}$; $y > 80$

7 a $\tfrac{1}{2}ah$, $\tfrac{1}{2}bh$. So $A = \tfrac{1}{2}(a+b)h$ b $h = \dfrac{2A}{a+b}$ c 4·5

8 a $x = 10-2Q$ b $t = \dfrac{u-v}{g}$ c $n = \dfrac{3kE}{9k-E}$
  d (1) $w = \dfrac{W(s-1)}{s}$ (2) $W = \dfrac{sw}{s-1}$

9 a $h = \dfrac{3V}{\pi r^2}$ b $r = \sqrt{\dfrac{3V}{\pi h}}$; $\sqrt{30} = 5\cdot48$ to 2 decimal places

10 a $20\tfrac{1}{4}$ b $132\tfrac{1}{4}$ c $12\tfrac{1}{4}$ d $90\tfrac{1}{4}$ e 5625

11 a $v = \dfrac{20u}{u+20}$ b $v = \dfrac{ur}{2u-r}$ 12 Both are true.

Answers

Algebra—Answers to Cumulative Revision Exercises

Page 76 Cumulative Revision Exercise A

1 a T b F c F d F
2 a {2, 4, 6, 8, 10, 12} b {1, 3, 5, 7, 9, 11} c {2, 3}
3 The set of the first six natural numbers. $n(S) = 6$. $A = \{2, 4, 6\}$.
   $B = \{1, 3, 5\}$. $A \cap B = \emptyset$ and $A \cup B = \{1, 2, 3, 4, 5, 6\}$
4 a $A \cap B = \{1, 2, 3\}$; $P \cap Q = \{4, 5, 6\}$; $R \cap S = \emptyset$.

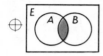

b $B \subset Q$ is false.

5 (i) 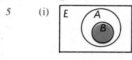 (ii)   (iii)

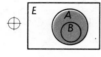

6 a {5, 6, 7, 8} b {1} c {5, 8} d {5, 8}
7 8   8 $P = \{8, 9, 10, \ldots\}$, $Q = \{5, 6, 7, 8, \ldots\}$, $P \cap Q = \{8, 9, 10, \ldots\}$
9

$B \cap C$ double-shaded;
$A \cap C =$ line segment HK

10a {2} b {−2} c {−1} d {−1, 0, 1}
  e {−2, −1} f {−2, −1, 1, 2, 3} g {1, 2} h {−1, 0, 1, 2, 3}
11a 0 b 2 c −2 d 5 e 7 f 2
12a 1 b 5 c −18 d −7a e −2b f 0
  g 63 h −48x^2 i 7 j −10 k −7 l −8
  m 6x^2 n 5m o 1 p 10 q −5b^2 r 6c
13a 5(p+q) b 4(n−2) c a(b+1) d a(a−4)
  e 3(2p+7q) f x(x+8) g 10x(1−2y) h 10x(1−2x)
14a 4p+2q or 2(2p+q) b −2q c 2q d 7x e 2x f a^2 g 25
15a 0 b 2x+6y−12z   16a (1) 2 (2) −10 (3) −24
  (4) $\frac{3}{2}$ b 1 c 1300 d −x e 0
17a 8p+7p+3p = 108; p = 6 b n+(n+1)+(n+2) = 225; 74, 75, 76
18a 9 b 3 c −5   19 14·4

Answers

20a {5, 6, 7}   *b* 29, 17   *21a* F   *b* F   *c* T   *d* T
22a 33·75   *b* 15   *23a* {11}   *b* {−7}   *c* {16}   *d* {$\frac{27}{4}$}   *e* {−1}
24a {2, 3, 4, ...}   *b* {4, 5, 6, ...}   *c* ø   *d* W   *e* {0, 1, 2, 3, 4, 5}
25a 6·5   *b* 12·5   *26a* 18 cm   *b* 6 cm
27a 56, 57   *b* 94, 96   *c* 863, 865
28

29   {(2, 2), (2, 6), (2, 8), (3, 3), (3, 6), (3, 9), (6, 6), (8, 8), (9, 9)}

30a T   *b* F   *c* F
31a   *b* {(1, 1), (2, 1), (3, 1), (3, 3)}

more than one arrow leaves some elements of A.

32

x	0	1	2	3	4
$2x-1$	−1	1	3	5	7

33

x	−4	−3	−2	−1	0	1	2	3	4
x^2	16	9	4	1	0	1	4	9	16

34a   *b*   *c*   *d*

35

36a {(1, 2)}   *b* {(3·5, −0·4)}
37a {(7, 4)}   *b* {(−1, −1)}   *c* {(2, 0)}

38a For x-axis, (3, 0); for y-axis (0, 3)   *b* I, IV, V   *c* III   *d* I, II
39a A(0, 8), B(5, 8), C(5, 0)   *b* interior of $\triangle$ AOC
   c {$(x, y): x \leq 5, y \leq 8, 8x+5y \geq 40$}

40a $x = n - m$ *b* $x = \frac{q}{p}$ *c* $x = \frac{c-b}{a}$
   d $x = \frac{d-2c}{2}$ *e* $x = \frac{b-a}{a}$ *f* $x = \frac{r}{p+q}$
   g $x = 2k$ *h* $x = ab$ *i* $x = \frac{mn}{p}$

41 $P = 4x+16$, $A = x^2+8x$; $18 < P < 20$ and $4\frac{1}{4} < A < 9$
42 $a = 3, b = -4$

Answers

Page 81 Cumulative Revision Exercise B

1 *a* T *b* F *c* F *d* T *e* T *f* F *g* T *h* T
2 *a* ∅ *b* W *c* T *d* S
3 *a* $A = \{2, 3, 5, 7\}, B = \{1, 3, 5, 7, 9\}, C = \{2, 3, 4, 6, 12\}$
  b $\{3\}, \{2, 3\}, \{3\}$ *c* $\{2\}$
4 *a* $\{3, 5\}$ *b* $\{2, 5\}$ *c* $\{5, 9\}$ *d* $\{2, 5, 9\}$ *e* $\{2, 5, 9\}$
  the distributive law of intersection over union
5 *a* 10 *b* 6   **6***a* T *b* F *c* F *d* T
7 *a* the set of people who play golf but not cricket or football
  b the set of people who play football and cricket but not golf
8 *a* F *b* T *c* F
9 *a* $\{\frac{3}{2}\}$ *b* $\{x : x < \frac{14}{3}\}$ *c* $\{x : x \leq 2\}$ *d* $\{12\}$ *e* $\{-1\}$
  f $\{x : x < \frac{7}{3}\}$
10*a* 2 *b* 30 *c* 0 *d* 40 *e* -10 *f* 116
11*a* $10x^2$ *b* $3(2a-b)$ *c* $-4q^2$ *d* $2x-3y$
  e $2a+3b$ *f* 1 *g* $4x^2$ *h* $-x^2$
12*a* $5y$ *b* $a-b$; $3x^2-2x$ **13***a* III *b* I *c* III *d* II
14 *a, b, c* are correct.
15*a* $2x+9$ *b* no *c* (1) 3 (2) -9 (3) -18 (4) 2
16*a* (1) 8 (2) 16 (3) 46 (4) 50 (5) -4 (6) -8 *b* no
17*a* $-2 < 5$ *b* $-7 < -4$ *c* $0 > -1$ *d* $-2 < 2$
18 $A = 100 - 4x < 80$, so $4x > 20$. Hence $x > 5$. $5 < x < 10$.
19*a* $2(2n+3)$ *b* $2h(a+b+c)$ *c* $4x(2x-1)$
20*a* (1) $(22-x)$ (2) $(18-x)$ *b* $(22-x)+x+(18-x) = 30 \Leftrightarrow x = 10$
21*a* $\{-5\}$ *b* $\{-\frac{11}{5}\}$ *c* $\{-1\}$ *d* $\{-1\}$
22*a* $\{x : x > 4\}$ *b* $\{x : x > -\frac{3}{4}\}$ *c* $\{x : x \geq 9\}$ *d* $\{x : x < \frac{15}{4}\}$
23*a* $P = 24x = 112$, so $x = \frac{14}{3}$ *b* $A = 22x^2 = 176$, so $x^2 = 8 \Rightarrow x = \sqrt{8}$
24*a* $\{x : x \geq 4\}$ *b* $\{x : -2 < x \leq 3\}$
25 area = 189 mm² $3a - 6 > 0 \Leftrightarrow 3a > 6 \Leftrightarrow a > 2$
26*a* $x < 100$ and $x > 48$ *b* 5 cm *c* $48 < 25\pi < 100 \Leftrightarrow 1.92 < \pi < 4$
27 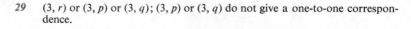   **28***a* F *b* T *c* T *d* F *e* T

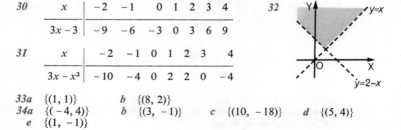

29 $(3, r)$ or $(3, p)$ or $(3, q)$; $(3, p)$ or $(3, q)$ do not give a one-to-one correspondence.

30

x	-2	-1	0	1	2	3	4
$3x-3$	-9	-6	-3	0	3	6	9

31

x	-2	-1	0	1	2	3	4
$3x-x^2$	-10	-4	0	2	2	0	-4

32 (graph shown)

33*a* $\{(1, 1)\}$ *b* $\{(8, 2)\}$
34*a* $\{(-4, 4)\}$ *b* $\{(3, -1)\}$ *c* $\{(10, -18)\}$ *d* $\{(5, 4)\}$
  e $\{(1, -1)\}$

Answers

35a $x = \dfrac{y-3}{6}$ b $x = \dfrac{y-c}{m}$ c $x = \pm\sqrt{(a+b)}$

d $x = \dfrac{ab}{c}$ e $x = \dfrac{bc-ad}{d}$ f $x = \dfrac{c}{a+b}$

36a $n = \dfrac{Ir}{E-IR}$ b $n = \pm \dfrac{1}{\sqrt{(a^2+1)}}$ c $n = \dfrac{t}{t-d}$

37   $\pi r = 2x$ 38a $r = \dfrac{3V + \pi d^3}{3\pi d^2}$ b 5; 4 cm

39   $h = \dfrac{A - 2\pi r^2}{2\pi r}$; $h = 3\tfrac{1}{2}$ 40 $R = \dfrac{r_1 r_2}{r_1 + r_2}$; 18

41a $p = -3$ and $q = -10$ b $P\left(\dfrac{b}{m-a}, \dfrac{mb}{m-a}\right)$ 42 $Q = 15\pi d^2 v$

43   $A' = \{x : x < 1\}$ $B' = \{x : x \geq 2 \text{ or } x \leq -2\}$

44a

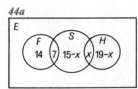

$14 + 7 + 15 - x + x + 19 - x = 50 \Leftrightarrow x = 5$
b   5   c   10

45a (1) $3x - 4$ (2) $-2a - 3$ (3) $-3x - 4y$ b $n = 3$

Geometry—Answers to Chapter 1

Page 89   Exercise 1

1   130°, 225°, 317°   2a 090°   b 180°   c 270°   d 045°

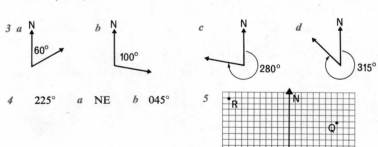

4   225°   a NE   b 045°

8   latitude and longitude; distance and bearing from well-known place.
9 a   bearing required   b 530 km, bearing 340°   10a no   b yes
11   angle of elevation of string

Answers

Page 92   Exercise 2

1

3 *a* Two sides of a triangle are together greater than the third.
   b $\vec{AB} \oplus \vec{BC} = \vec{AC}$
4 *a* $\vec{PR} \oplus \vec{RQ} = \vec{PQ}$ *b* $\vec{PQ} \oplus \vec{QR} = \vec{PR}$ *c* $\vec{RP} \oplus \vec{PQ} = \vec{RQ}$
5 *a* $\vec{VU} \oplus \vec{UT} = \vec{VT}$ *b* $\vec{UV} \oplus \vec{VT} = \vec{UT}$ *c* $\vec{UT} \oplus \vec{TV} = \vec{UV}$
6 $\vec{PQ} \oplus \vec{QR} = \vec{PR}$; $\vec{PS} \oplus \vec{SR} = \vec{PR}$
7 *a* $\vec{BA}$ *b* $\vec{AB} \oplus \vec{BA} = 0$ *c* no
8 *a* Car returns to starting-place.
   b Sum of three non-zero magnitudes cannot be zero.
9 *a* any quadrilateral ABCD
   b $\vec{AB} \oplus \vec{BC} = \vec{AC}$, $\vec{AB} \oplus \vec{BC} \oplus \vec{CD} \oplus \vec{DA} = 0$, etc

Page 93   Exercise 3

1 *a* 8 m, B′   *b* 8 m, C′   *c* 8 m, $\vec{AA'}$
2 *a* (1) and (2) 8 paces forward   *c* same shape and size
3 *a, b, c* 300 m   **4** its direction   **5** *a* 2 cm   *b* parallel to BC
6 *a* 2 cm   *b* parallel to AL   **7** *a* 2, 12, 16, 19   *b* B, C, D, E
8 *a* 10, 16, 19, 13   *b* G, H, K, L

Page 95   Exercise 4

1 *a* 6, 6, 4 cm   *b* AD ∥ BC, AB ∥ DC
2 *a* 8 cm   *b* 8 cm   *c* 6 cm   *d* 7 cm
3 *a* PK ∥ QM ∥ RN   *b* PK = QM = RN   *c* △KMN, congruent
   d PKMQ, QMNR, PKNR   **4***b* $\vec{BX}, \vec{CY}, \vec{DZ}$ **5** $\vec{CA}$
7 *a* $\vec{DF}$, etc   *b* $\vec{FD}$, etc   **8***a* T   *b* F   *c* T   *d* F   *e* T   *f* T

Page 97   Exercise 4B

1 *a* AFE, FBD, EDC, DEF   *b* FBD, $\vec{FA}$, etc.; EDC, $\vec{EA}$, etc.
   c DEF, half turn
2 $\vec{DC}, \vec{FE}$; $\vec{DE}$ or $\vec{FA}$ or $\vec{BF}$ **3** QQ′, RR′, SS′; 6
4 plane completely covered without overlapping

Page 98   Exercise 5

1 all equal in magnitude and direction   **2** D, H, L
3 (3, 3), (4, 6), (8, 4), (−3, 3), (0, 2)   **4** $\binom{1}{2}, \binom{0}{-2}, \binom{2}{-2}, \binom{3}{0}, \binom{-4}{-1}$

Answers

5   (2, −1), (7, 6), (0, −3), (3, 0), (12, 9)

6 a $\begin{pmatrix} -1 \\ 4 \end{pmatrix}$ b   (0, 6), (5, 10), (0, 0)

8   (11, 5), (9, 9), (3, 9), (2, 2), (6, 0); (6, 5)

9 a   (9, 2), (5, 0)   b $\vec{AA'} = \vec{BB'}$ or $\vec{AB} = \vec{A'B'}$, ABB'A'

10a   (9, 1), (−5, −2)   b $\vec{AA'} = \vec{BB'}, \vec{AB} = \vec{A'B'}$

Page 100   Exercise 5B

1 a $\begin{pmatrix} 2 \\ 0 \end{pmatrix}$; (2, 6), (1, 0), (3, 0)   b $\begin{pmatrix} -2 \\ 0 \end{pmatrix}$; (−2, 6), (−3, 0), (−1, 0)

   c $\begin{pmatrix} 0 \\ 6 \end{pmatrix}$; (0, 12), (−1, 6), (1, 6)

2    $\begin{pmatrix} 7 \\ -1 \end{pmatrix}$, (7, −1). They are parallel and equal.

3   (9, 6); (6, 4), (12, 4), (12, 8), (6, 8)

4 a $\begin{pmatrix} 7 \\ 3 \end{pmatrix}$ b $\begin{pmatrix} -2 \\ 7 \end{pmatrix}$ c $\begin{pmatrix} 6 \\ -4 \end{pmatrix}$ d $\begin{pmatrix} -6 \\ -5 \end{pmatrix}$ e $\begin{pmatrix} 0 \\ 4 \end{pmatrix}$ f $\begin{pmatrix} -6 \\ 0 \end{pmatrix}$

5 a   (1) $\begin{pmatrix} 6 \\ 4 \end{pmatrix}$ (2) $\begin{pmatrix} 9 \\ 6 \end{pmatrix}$ (3) $\begin{pmatrix} 30 \\ 20 \end{pmatrix}$ (4) $\begin{pmatrix} -15 \\ -10 \end{pmatrix}$ b $\begin{pmatrix} 2a \\ 2b \end{pmatrix}, \begin{pmatrix} 3a \\ 3b \end{pmatrix}, \begin{pmatrix} 10a \\ 10b \end{pmatrix}, \begin{pmatrix} -5a \\ -5b \end{pmatrix}$

   c   Multiply each component by the number.

Page 101   Exercise 6

1   see Figure 16

2 a   (1) Y   (2) Z   (3) A   (4) C   b   yes   c   yes, yes

4   b, c, g, h   5   c, d, e, f, g

Page 104   Exercise 7

1 a $\begin{pmatrix} 2 \\ 0 \end{pmatrix}, \begin{pmatrix} 0 \\ 4 \end{pmatrix}$ b $\begin{pmatrix} 3 \\ 0 \end{pmatrix}, \begin{pmatrix} 0 \\ 2 \end{pmatrix}$ c $\begin{pmatrix} -2 \\ 0 \end{pmatrix}, \begin{pmatrix} 0 \\ 4 \end{pmatrix}$ d $\begin{pmatrix} 3 \\ 0 \end{pmatrix}, \begin{pmatrix} 0 \\ -5 \end{pmatrix}$ e $\begin{pmatrix} 4 \\ 0 \end{pmatrix}, \begin{pmatrix} 0 \\ -2 \end{pmatrix}$

2    $\begin{pmatrix} 2 \\ 0 \end{pmatrix} + \begin{pmatrix} 0 \\ 4 \end{pmatrix} = \begin{pmatrix} 2 \\ 4 \end{pmatrix}, \begin{pmatrix} 3 \\ 0 \end{pmatrix} + \begin{pmatrix} 0 \\ 2 \end{pmatrix} = \begin{pmatrix} 3 \\ 2 \end{pmatrix}, \begin{pmatrix} -2 \\ 0 \end{pmatrix} + \begin{pmatrix} 0 \\ 4 \end{pmatrix} = \begin{pmatrix} -2 \\ 4 \end{pmatrix},$

     $\begin{pmatrix} 3 \\ 0 \end{pmatrix} + \begin{pmatrix} 0 \\ -5 \end{pmatrix} = \begin{pmatrix} 3 \\ -5 \end{pmatrix}, \begin{pmatrix} 4 \\ 0 \end{pmatrix} + \begin{pmatrix} 0 \\ -2 \end{pmatrix} = \begin{pmatrix} 4 \\ -2 \end{pmatrix}$

3 a $\vec{DE} \oplus \vec{EF} = \vec{DF}, \begin{pmatrix} 3 \\ 2 \end{pmatrix} + \begin{pmatrix} -2 \\ 4 \end{pmatrix} = \begin{pmatrix} 1 \\ 6 \end{pmatrix}$

   b $\vec{PQ} \oplus \vec{QR} = \vec{PR}, \begin{pmatrix} -3 \\ -3 \end{pmatrix} + \begin{pmatrix} 6 \\ -2 \end{pmatrix} = \begin{pmatrix} 3 \\ -5 \end{pmatrix}$

   c $\vec{LM} \oplus \vec{MN} = \vec{LN}, \begin{pmatrix} 4 \\ -2 \end{pmatrix} + \begin{pmatrix} 2 \\ 6 \end{pmatrix} = \begin{pmatrix} 6 \\ 4 \end{pmatrix}$

   d $\vec{GH} \oplus \vec{HK} = \vec{GK}, \begin{pmatrix} 3 \\ 3 \end{pmatrix} + \begin{pmatrix} 2 \\ -5 \end{pmatrix} = \begin{pmatrix} 5 \\ -2 \end{pmatrix}$

4 b $\vec{UV} \oplus \vec{VW} = \vec{UW}$ c $\begin{pmatrix} -2 \\ 4 \end{pmatrix} + \begin{pmatrix} 3 \\ -1 \end{pmatrix} = \begin{pmatrix} 1 \\ 3 \end{pmatrix}$

5 a $\begin{pmatrix} -1 \\ 5 \end{pmatrix}$ b $\begin{pmatrix} 0 \\ 0 \end{pmatrix}$ c $\begin{pmatrix} 11 \\ -11 \end{pmatrix}$ d $\begin{pmatrix} p+r \\ q+s \end{pmatrix}$ 6   2, 4

9   63 km, 031°; $\vec{AC} \oplus \vec{CB} = \vec{AB}$ 10   start and finish at same point

Answers

Page 106   Exercise 8

4 *a* 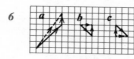   *b* $\binom{4}{4}+\binom{2}{-2}=\binom{6}{2}$

5 *b* (10, 4), (8, −2), (12, 2); (14, 5), (12, −1), (16, 3)

6 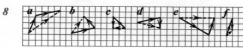   **7***a* zero answer   *b* $\binom{2}{-2}+\binom{-2}{2}=\binom{0}{0}$

8

9 *a* (6, 6), (5, 0)   *b* $\binom{7}{0}$ **10** $\binom{0}{0}$ *a* $\binom{-2}{-3}$ *b* $\binom{5}{-1}$ *c* $\binom{0}{0}$

Page 108   Exercise 8B

1 parallelogram   **2***b* $\binom{-2}{-6}+\binom{-3}{-9}=\binom{-5}{-15}$

3 $\binom{-2}{-6}+\binom{3}{9}=\binom{1}{3}$ **4***a* $\binom{5}{-1}$ *b* $\binom{-3}{-5}$ *c* $\binom{2}{8}$

5 commutative, associative

Geometry—Answers to Chapter 2

Page 110   Exercise 1

1   1, 4, 9, 16, 25
2 *a* 6   *b* 8   *c* 10   *d* 12   *e* 4·47   *f* 7·42   *g* 8·37   *h* 2·45
3 *a* 16   *b* 8   *c* 8   *d* 2·83   **4***a* 49   *b* 24   *c* 25   *d* 5
5   4 right, 3 up; 3 left, 4 up; 4 left, 3 down; 3 right, 4 down
6 *b* 5 right, 2 up   *c* square   **7***b* 3 right, 5 up   *c* square
8 *c* 13, 3·61   **9***a* 100, 10   *b* 5, 2·24   *c* 41, 6·40
10*a* 8, 25, 13, 100, 5, 41   *b* same numbers in two rows

Page 113   Exercise 2

1   (i) $a^2 = b^2+c^2$ (ii) $f^2 = d^2+e^2$ (iii) $q^2 = p^2+r^2$ (iv) $x^2 = y^2+z^2$
2   (i) 5   (ii) 13   (iii) 17   (iv) 25   **3***a* $b^2 = a^2-c^2$, $c^2 = a^2-b^2$
   b (i) 6   (ii) 9   (iii) 4·47   (iv) 6·24   **4**   10, 3·61; 12, 9·17; 8, 6·71; 50, 200
5   7·8 cm   **6**   6·6 cm   **7**   9·54 m   **8**   9 m   **9**   8·5, 5 cm
10   12·6 cm   **11**   8 km   **12***a* square   *b* 7·07 cm   *c* 8·66 cm
13*a* rectangle   *b* 10 cm   *c* 10·8 cm

Answers

Page 115 Exercise 2B

1. 12 cm, 60 cm² 2 5·66 3a 24 cm² b 10 cm c 4·8 cm
4 a 30 cm² b 13 cm c 4·62 cm 5 170 km
6 12 cm, 10·9 cm 7 10·9 cm 8 17 cm, 18·8 cm 9 $s^2 = a^2 + b^2 + c^2$

Page 117 Exercise 3

1 a 4, 3 b 5 2a 2, 1 b 2·24 3a 4, 2 b 4·47
4 a 3, 5 b 5·83

Page 119 Exercise 4

1 a 5 b 2·24 c 4·47 d 5·83
2 a 5 b 10 c 13 d 13 e 15 f 4·24
3 AB = $\sqrt{37}$ = AC
4 each distance from origin = 5 5 *a, c, d*
6 b 1·41, 4·24, 4·47 c rectangle

Page 119 Exercise 4B

1 a 4·12, 3·61; 7·21, 2·83; parallelogram b 4·47; 8·49, 2·83; rhombus
2 each distance from (2, 2) = 5 3a 5p b 7·28m
4 6·7 km 5 13, distance from centre = 13 6 6·32, (±5·57, 0)

Page 121 Exercise 5

1 T   2 F   3 F   4 T   5 T
6 F   7 F   8 F   9 T   10 F

Page 122 Exercise 6

1 *a, d, f, g, h* 2a 15 b 20 4 rhombus

Page 123 Exercise 6B

1 *b, c, d, e, f, h* 4 15, 20, 21·9 5 $\sqrt{65}$, 5, $\sqrt{40}$; angle QSR
6 $y^2 - z^2 = a^2$; sides *a* and *z*

Geometry—Answers to Revision Exercises

Page 128 Revision Exercise 1

2 b 10 km, 053°; 17 km, 118°; 13 km, 337°
3 when he has flown the required distance
4 94 km; 112° 5a 13 km, 293° b 12 km east, 5 km south
6 a $\overrightarrow{BC}$ b $\overrightarrow{BC}$ c $\overrightarrow{AD}, \overrightarrow{CD}$ d $\overrightarrow{CA}$
7 a $\overrightarrow{DB}$ b $\overrightarrow{DC}$ c $\overrightarrow{EA}$ d $\overrightarrow{CE}$ e $\overrightarrow{CB}$ f $\overrightarrow{EB}$
8 a 0 b 0 The sum of the segments given by the sides taken in order is zero.
11a a parallelogram b (8, 0), (12, 2), (14, 7), (10, 5)
   c (−6, −7), (−2, −5), (−4, −2)
12a F b F c T d F e T f T g (1) T (2) T
   (3) F (4) F (5) T   14b (1) $\begin{pmatrix} 8 \\ -4 \end{pmatrix}$ (2) $\begin{pmatrix} -8 \\ 4 \end{pmatrix}$

Answers

15a $\binom{2}{4}, \binom{1}{2}$ b (4, 4), (12, 12)   c M is always the midpoint of FG.

16 $\binom{2}{0}$ and $\binom{-10}{0}$

17a an isosceles triangle   b $\binom{-3}{0}$ c $\binom{0}{-8}, \binom{0}{10}$. Images of A are $(-2, -12)$ and $(-2, 6)$; images of B are $(8, -12)$ and $(8, 6)$.

18a $\binom{-7}{3}$ b $(-3, 4)$ c $\binom{-4}{-1}$, $(3, -4)$

19 On your diagram the required segments represent translations of:
  a $\binom{5}{1}$ b $\binom{6}{4}$ c $\binom{1}{5}$ d $\binom{5}{-1}$ e $\binom{0}{0}$ f $\binom{1}{3}$

20a $\binom{1}{-5}$ b $\binom{2}{-3}$ c $\binom{-1}{-4}$ d $\binom{1}{2}$ e $\binom{0}{0}$

21a (1) $\binom{0}{0}$ (2) $\binom{0}{0}$ (3) $\binom{0}{0}$ b since $\binom{a}{b} + \binom{-a}{-b} = \binom{0}{0}$

22a (1) $\binom{2}{3}$ (2) $\binom{4}{-7}$ (3) $\binom{-4}{7}$ (4) $\binom{-2}{-3}$
  b (4) is the negative of (1)   (3) is the negative of (2).

Page 132   Revision Exercise 2A

1 a 5   b 9   c 11   d 5·48   e 8·06   f 13·4   2 45, 6·71
3   (i) 10   (ii) 26   (iii) 30   (iv) 9
4   (i) 7·81   (ii) 11·7   (iii) 10·5   (iv) 8·66
5   12 cm   6   10 cm
7 a 15·6, 12·8, 14·4 cm   b rectangle   c 17·5 cm   8 1·68 m
9   6·40 m   10 10·9 km   11b AD   12b 0·81
13   7·62; inside   14 (5, 5); 6·71, 4·47; 8·06   15 5, 10
16   all three   19 11·3, 12·6 cm

Page 134   Revision Exercise 2B

1   16·2 km   2 11; 330, 10·8   3 72 m   4 85 m   5 25·1 m
6   9·7 m   7 8·7, 12·2, 5·8 m
8   14·1 km   a 141 km   b 1410 km; not sides of a triangle
9   14·1 cm, 1·41   a 2·24   b 1·73   10 rough road
11   (5, 5); 2·8, 11·3; 15·8   12a 13p   b 17m
13   each = 13; no   14 a, c, e   15 CA = CB, $(\frac{1}{2}, -\frac{5}{2})$
16a 8·66 cm   b 22·4, 21·8 cm   17 7·1, 8·6, 8·6
18   26·2, 23·8, 22·5, 22·0, 22·3, 24; 21·9

Geometry—Answers to Cumulative Revision Exercises

Page 146   Cumulative Revision Exercise A

1 a F   b T   c T   d T   2 cube, pyramid, cylinder, cuboid
3   3 cm   4 108 cm   7 H, I, O, X   8 360°
9 a 90°   b 180°   c 120°   d 15°
10a $\angle$ADE, $\angle$EAB, etc.
  b $\angle$DAB, $\angle$ABC, $\angle$AEB, $\angle$DEC
  c $\angle$AEB, $\angle$DEC; $\angle$AED, $\angle$BEC
  d $\angle$AED, $\angle$AEB; $\angle$AEB, $\angle$BEC; etc.
11 $\angle$SOP and $\angle$POQ, etc.; $\angle$PQR and $\angle$SRQ, etc.

Answers

12a supplementary   b neither   c complementary
 d neither   e supplementary   f complementary
13a T b F c F d T e F   14 QR = 4·9 cm, RP = 6·0 cm
15a F b F c T d T   16b $\frac{QP}{RS}$, $\frac{SR}{PQ}$ c $\frac{RS}{QP}$
17 (0, 0), (6, 0), (6, 4), (0, 4); (0, 0), (4, 0), (4, 6), (0, 6); (3, 2), (2, 3)
18 7·5 cm²   19a (i) 30 cm² (ii) 16 cm²   b (i) 8 cm (ii) 3, 5 cm
20 50°   22 $A_1(4, -4)$, $B_1(4, -2)$, $A_2(-4, 4)$, $B_2(-4, 2)$; each = 4 units²
23a (-3, -2)   b (3, -2)   c (5, 2)   d (2, -3)
24a T b F c T d F e F f T
25a (3, 2)   b (-3, -4)
26a PR = 6·7 cm, QS = 10·2 cm   b 31·7 cm²
27

	Rectangle	Rhombus	Square	Kite	Parm.	Isos. △	Equil. △
No. of ways	4	4	8	2	2	2	6
No. of axes	2	2	4	1	0	1	3

28a ∠BCD, ∠CDE   b ∠ADE, ∠ABC or ∠AED, ∠ACB

29

30

31

32 rectangle

33

34 b, d, e   35 58 km, 344°   36 (14, -12)

37

38 $O_1(6, 6)$, $A_1(12, 12)$, $B_1(12, 9)$; 9 square units

39 (1, -2), kite   40a 5   b 13   c 2·83   d $\sqrt{(p^2+q^2)}$
41a 15 cm   b 26 cm   c 29 cm   d 11·4 cm
42a 8 cm   b 40 cm   c 9·75 cm
43 75 km   44a 6·7   b 2·2   45 9·8 km

Page 152 Cumulative Revision Exercise B

1 4 cm   2 x = 4, kite   3a 45°   b 135°   c 130°
4 a 81°, 44°, 55°   b 1 cm < RP < 12 cm   5 51 m   6 18·3 m
7 a (2, 5)   b (3·5, 3)   c (2)   8 4 cm   9 (3, 10)
11b (2, 8)   c (0, 8), (4, 8)   d 6 units²   e rhombus
12a $A_1(0, 3)$, $B_1(-1, 2)$, $C_1(-4, 6)$ b $A_2(0, -3)$, $B_2(1, -2)$, $C_2(4, -6)$
  c $\begin{pmatrix} -2 \\ -6 \end{pmatrix}$, no   13b (2, 12)   c 24 square units   d y = 6

Answers

14c A(1, 6), B(3, 9), C(5, 6)   d 16 square units   e $B_1(3, 3)$, $D_1(3, 11)$
15b (−4, 11)   c (−1, 7)
16b parallelogram, 16 square units   c P(11, 1), Q(9, −3); parallelogram
 d (5, 5), (1, 5)
17a $\angle$ADE and $\angle$AGH, etc   b $\angle$ABD and $\angle$BDE, etc   c 2   d 6
18 rectangle   19b $\{(x, y): x \geqslant 0, y \leqslant 5, y \geqslant x\}$
 c rt-angled $\triangle$, $12\frac{1}{2}$ square units

20a 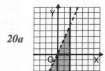   b

21a same direction, at 45° to x-axis   b each = 7·07
22a $p = 3, q = -4$ b $p = -2, q = -3$ 23 c
24 31·5 cm or 16·5 cm   25 32 km
26a $A_1(9, 14)$, $A_2(12, 14)$ b 7·81, 3, 10   c $a = 8, b = 6$
29b angles are all 30°, 60°, or 90°   c 10·4 cm, 31·2 cm²
30a 60 cm   b 660 cm²   c 21·6 cm   31 ±12, direction
32a AM = MB, DN = NC, AD = BC
 b $\angle$AMN, $\angle$BMN, $\angle$DNM, $\angle$CNM
 c $\angle$A = $\angle$B, $\angle$D = $\angle$C   d AB and DC
 e $\angle$A and $\angle$D, $\angle$A and $\angle$C, $\angle$B and $\angle$C, $\angle$B and $\angle$D, angles at M, etc
33a 2, −5   b 9   c 5, −15   d −2, −8

Arithmetic—Answers to Chapter 1

Page 161   Exercise 1

1   0, 1, 4, 9, 16, 25, 36, 49, 64, 81, 100
2   121, 144, 169, 196, 225, 256, 289, 324, 361, 400   3 625, 900, 2500, 5625
4   2·25, 20·25, 56·25, 90·25   5 0·01, 0·09, 0·25, 0·49, 0·81
6   4, 400, 40000   7 10000, 90000, 160000, 250000
8   16 cm², 49 m², 4 km², 4·41 m², 0·64 km², 22·09 cm²

Page 163   Exercise 2

1 a	64	b	81	c	20·3	d	5·3	e	50·4	f	42·3
g	49	h	100	i	10·9	j	31·4	k	67·2	l	90·3

Page 163   Exercise 3

1	6·25	2	12·25	3	20·25	4	20·34	5	20·43	6	21·07
7	49	8	49·14	9	50·41	10	84·64	11	86·30	12	48·86
13	4·08	14	29·92	15	76·74						

Page 164   Exercise 4B

1 a	552	b	1376	c	7957	d	102
2 a	207900	b	43700	c	250000	d	819000
3 a	0·0841	b	0·7569	c	0·3136	d	0·1089
4 a	166	b	23100	c	0·6084	d	0·6225
5 a	942	b	15400	c	426400	d	0·0424

Answers

Page 165 Exercise 5

1 a	2·25	b	5·76	c	9·61	d	2·40	e	6·00
2 a	20·3	b	31·4	c	60·8	d	89·3	e	38·7
3 a	4·49	b	14·1	c	3·35	d	68·1	e	82·8

Page 165 Exercise 5B

1 a	1·21	b	121	c	39·7	d	3970	e	397000
2 a	7·29	b	729	c	72900	d	0·0729	e	4·28
3 a	3·00	b	0·706	c	5450	d	9·30	e	15100

Page 165 Exercise 6

1 a	25	b	225	c	625	d	2500	e	250000
f	72·25	g	0·25	h	156	i	289	j	5041
k	0·1521	l	0·0605	m	16400	n	74000	o	250
2 a	12·25 cm²	b	256 mm²	c	1·12 m²	d	1376 cm²		
3 a	178	b	8·18	c	68				

Page 166 Exercise 7

1 a	1	b	4	c	6	d	8	e	10	f	20
2 a	5	b	7	c	9	d	12	e	30	f	40

3 $\sqrt{289} = 17$ 4 $\sqrt{529} = 23$ 5 $\sqrt{12·25} = 3·5$
6 $\sqrt{1000000} = 1000$

7 a	3 cm	b	6 m	c	10 km	d	15 cm	e	1·2 m

Page 167 Exercise 8

1 a	4·5	b	5·5	c	6·3	d	7·1	e	7·8	f	8·4
2 a	3·9	b	7·4	c	8·7	d	9·4	e	9·5	f	6·1
3 a	4·4 cm	b	7·4 m	c	8·6 mm	d	9·8 m	e	5·3 cm	f	8·9 mm

Page 167 Exercise 9

1 a	2–3	b	4–5	c	9–10	d	11–12	e	3–4	f	6–7
2 a	6·5	b	8·4	c	9·7	d	2·8	e	2·2	f	5·5
3 a	3·5	b	7·1	c	1·7	d	2·7	e	9·5	f	2·5

4 5·2, 1·6   5a 2·5, 7·8   b 6·9, 2·2   c 9·8, 3·1   d 1·2, 3·7

Page 168 Exercise 10

1 a	2·24	b	2·24	c	2·25	d	2·26	e	2·26	f	2·28
g	2·61	h	2·77	i	3·01	j	1·00	k	1·87	l	2·61
2 a	3·16	b	5·48	c	5·53	d	6·00	e	7·27	f	9·36
3 a	8·77	b	4·71	c	1·41	d	1·95	e	7·41	f	2·53
4 a	7·07 cm	b	4·04 mm	c	9·38 cm	d	1·32 m	e	2·66 mm		
5 a	3·74	b	12	c	1·72						

Page 170 Exercise 11B

1 a $2·34 \times 100$ b $6·38 \times 100$ c $2·00 \times 100$ d $21·35 \times 100$
  e $30·47 \times 100$

2 a $\dfrac{52}{100}$ b $\dfrac{63}{100}$ c $\dfrac{50}{100}$ d $\dfrac{14·5}{100}$ e $\dfrac{20·4}{100}$

Answers

3 a $\dfrac{6}{100}$ b $\dfrac{2\cdot 5}{100}$ c $\dfrac{70}{100}$ d $\dfrac{2\cdot 46}{100}$ e $\dfrac{77\cdot 77}{100}$

4 a	11·6	b	29·5	c	20·2	d	22·4	e	23·8
5 a	35·1	b	76·8	c	93·8	d	49·9	e	16·7
6 a	0·6	b	0·938	c	0·888	d	0·671	e	0·212
7 a	0·270	b	0·134	c	20·8	d	0·958	e	61·1
8 a	80·6	b	1·53	c	28·4	d	6·07	e	0·868
f	0·245	g	48·0	h	3·54	i	0·987	j	22·3
9 a	116	b	157	c	351	d	366	e	260
10a	0·0894	b	0·0872	c	0·0224	d	0·0283	e	0·0539

Page 171 Exercise 12

1 a	1·73	b	2·83	c	5·48	d	8·25	e	9·75	f	6·93
2 a	2·10	b	2·76	c	3·13	d	5·95	e	9·10	f	2·31
3 a	13·5	b	26·0	c	53·8	d	86·5	e	29·2	f	6·72
4 a	0·141	b	0·447	c	0·883	d	0·279	e	0·620	f	0·196
5 a	13·9	b	0·823	c	93·6	d	0·311	e	0·675	f	20·8

Page 173 Exercise 13

1 a	2·45	b	5·39	c	9·38	d	7·48	e	6·32
2 a	2·65	b	9·49	c	4·12	d	4·47	e	1·73
3 a	3·85	b	2·86	c	8·68	d	7·53	e	1·26
4 a	3·4641	b	8·3066	c	2·8284	d	4·6476	e	9·6644

Arithmetic—Answers to Chapter 2

Page 179 Exercise 3B

3 a	6	b	15	c	16	d	14	e	15
f	6	g	12	h	6·25	i	3	j	2
k	3	l	2	m	1·6	n	4	o	4

Page 180 Exercise 4

1 Figure 8: 1·1; 1·25; 1·4; 1·6; 1·74. Figure 9: 3·2; 3·32; 3·66; 3·85. Figure 10: 7·05; 7·2; 7·45; 7·78
2 3·2; 6·2; 9·0; 2·48; 4·44; 9·8. multiplying by 2

Page 181 Exercise 5

1	6·00	2	10·0	3	7·50	4	3·60	5	10·0	6	6·40
7	10·0	8	9·00	9	3·50	10	9·00	11	4·84	12	9·86

Page 182 Exercise 6

1	2·00	2	2·00	3	3·50	4	2·50	5	4·00	6	1·33
7	1·14	8	1·60	9	1·67	10	1·40	11	1·57	12	1·05

Page 182 Exercise 7

1	5·48	2	8·12	3	1·73	4	1·24	5	4·47	6	9·38
7	81·0	8	13·0	9	26·0	10	51·8	11	3·61	12	1·17

Answers

Page 183   Exercise 8

1	1000	*2*	80	*3*	20	*4*	0·9	*5*	0·02
6	900	*7*	6	*8*	10	*9*	0·03	*10*	0·07
11	40	*12*	100	*13*	400	*14*	6000	*15*	20 cm²
16	0·04 cm²	*17*	7 m³	*18*	800 m	*19a*	8000 g	*b*	20 000 g
20	40 000 km; 10 000 km								

Page 184   Exercise 9

1	200; 209	*2*	200; 234	*3*	400; 378	*4*	90; 80·0
5	800; 779	*6*	8; 7·77	*7*	6; 6·53	*8*	6; 6·58
9	700; 757	*10*	80; 92·2	*11*	0·8; 0·936	*12*	50 000; 56 100
13	0·8; 0·806	*14*	0·07; 0·0942	*15*	0·01; 0·00978		

Page 184   Exercise 10

1	20; 14·6	*2*	20; 21·1	*3*	20; 18·0	*4*	4; 3·39
5	20; 20·2	*6*	2; 1·89	*7*	20; 22·7	*8*	20; 20·1
9	0·2; 0·245	*10*	2; 2·26	*11*	0·001; 0·00113	*12*	30; 29·7

Page 185   Exercise 11

1	35·0	*2*	6470	*3*	47·0	*4*	0·622	*5*	261
6	7·88	*7*	4·42	*8*	0·118	*9*	0·0528	*10*	0·500
11	3·00	*12*	9·46	*13*	0·0353	*14*	0·0755	*15*	0·0635
16	0·0382	*17*	892	*18*	0·00882				

Page 186   Exercise 12

1	5·45	*2*	12·2	*3*	10·5 cm²	*4a*	5·39	*b*	40·6		
5 a	57·1%	*b*	62·5%	*c*	88·9%	*d*	15·6%	*e*	56·2%	
6	4·59	*7*	0·579	*8*	77·4 cm²	*9*	4·30 m			
10	$1910	*11*	£892	*12*	£830	*13*	21·9			
14	1·51	*15*	1·57 m³	*16*	0·500 cm	*17*	561 km/h			
18	6 h 43 min	*19*	8·91 kg	*20*	535	*21*	2·28			

Page 187   Exercise 8 (by slide rule)

1	1070	*2*	93·4	*3*	15·2	*4*	0·609
5	0·0230	*6*	882	*7*	6·02	*8*	14·4
9	0·0257	*10*	0·0696	*11*	47·4	*12*	166
13	396	*14*	4790	*15*	16·2 cm²	*16*	0·0529 cm²
17	7·48 m³	*18*	770 m	*19a*	8260 g	*b*	21 600 g
20	39 800 km; 10 900 km							

Arithmetic—Answers to Chapter 3

Note. Where a slide rule has been used, the answers obtained may not be exactly the same as those given.

Page 193   Exercise 2

1 a	22 cm	*b*	66 cm	*c*	110 cm	*d*	154 cm
	e	31·4 m	*f*	12·6 cm	*g*	25·1 mm	*h*	7·54 m
2 a	88 cm	*b*	132 cm	*c*	176 cm	*d*	352 cm
	e	12·6 m	*f*	62·8 m	*g*	31·4 m	*h*	50·9 m

Answers

3   62·8 mm, or 6·28 cm   *4*   88 cm   *5*   94·2 cm
6 a   6·28 cm   *b*   75·4 cm   *7*   18·8 m   *8*   19·8 m   *9*   126 m
10   40 000 km   *11a*   132 cm   *b*   66 m   *12*   420 m   *13*(i) 25·7 m   (ii) 14·3 m
14a   14 cm   *b*   17·5 m   *c*   35 mm   *d*   4·78 m
15a   21 cm   *b*   10·5 cm   *16*   63·6 m   *17*   35 m   *18*   64 cm
19   9·55 m   *20*   3·18 cm

Page 195   Exercise 2B

1 a   (*1*) 198 cm   (*2*) 11 cm   (*3*) 314 m
   b   (*1*) 528 mm   (*2*) 66 cm   (*3*) 18·8 km
2   11 km   *3*   66 m   *4*   80 m   *5a*   60°, $\frac{1}{6}$ *b* $\frac{1}{6}$ *c*   1·05 cm
6 a   88 cm   *b*   (*1*) 22 cm   (*2*) 11 cm   (*3*) 29·3 cm   (*4*) 66 cm   *7* 176 cm, 250
8 a   25 m   *b*   35 m   *c*   100 m   *d*   239 m   *9a*   5 m   *b*   79·6 cm
10   24 200 km/h   *11*   6100 km   *12* $9·4 \times 10^8$ km

Page 198   Exercise 4

1 a   154 cm²   *b*   616 cm²   *c*   314 cm²   *d*   12·6 cm²
2 a   38·5 mm²   *b*   3·14 cm²   *c*   78·5 m²   *d*   0·785 km²
3   314 mm²   *4*   1390 m²   *5*   3·14 cm²   *6*   62·4 m²
7   706 cm²   *8*   10 800 m²   *9*   (i) 39·2 m²   (iii) 12·6 m²
10   (i) 3·22 m²   (ii) 14·3 m²
11a   10 cm   *b*   7 cm   *c*   2·65 cm   *d*   6·26 cm
12   10·6 m   *13*   14 m   *14*   4·89 m   *15a*   20 cm   *b*   126 cm

Page 200   Exercise 4B

1 a   (*1*) 38·5 cm²   (*2*) 31 400 m²   *b*   (*1*) 9·62 cm²   (*2*) 50·2 m²
2   113 cm²   *3*   23·4 m²   *4*   21·5%   *5*(i) 7·7 cm²   (ii) 42 m²
6 a   3·74 m   *b*   12·4 cm   *c*   17·8 m   *7*   1·60 cm   *8*   14·1 m
9 a   2:1   *b*   4:1   *10*   2, 1; a rectangle; length = circumference;
     44 cm, 5 cm; 220 cm²   *a*   126 cm²   *b*   113 cm²   *c*   70·4 cm²

Arithmetic—Answers to Chapter 4

Page 203   Exercise 1

1 a   1958   *b*   1954, 225 000   *c*   29%
3 a   3643   *b*   2·2%

Page 206   Exercise 2

1   1   *2*   37   *3*   69

Page 208   Exercise 3

2   37–39   *3*   69–70

Page 210   Exercise 4

1 $\frac{11}{23}$ *3* $\frac{1}{10}$

Page 212   Exercise 5

4 $\frac{7}{22}$ *5a*   120   *b*   3   *c* $\frac{3}{10}, \frac{17}{30}$ *6a* $\frac{1}{50}$ *b* $\frac{17}{50}$

Answers

Page 216   Exercise 6

1 a	9	*b*	19 cm	*c*	18·0 kg	*d*	95p	*2*	7·1 minutes
3	4·9 hours	*4a*	5·9	*b*	7·2	*5a*	£181 million	*b*	£211 million
7 a	4; 4; 4	*b*	£13; £14·22; £14	*c*	2; 4; 3·5	*8*	70	*9*	mode

Page 217   Exercise 6B

1	24·5	*2*	5·8	*3*	50·5 kg	*4a*	247 000	*b*	274 000
5	£4000	*6*	162 cm	*7*	9·12 km; 255 000 m²			*8*	2·1; 2; 1
9	9·2 km/litre	*10*	42 kg	*11*	£71 million				

Page 219   Exercise 7

1	1·6	*2*	16·7	*3*	5·7	*4*	6·4	*a*	29%	*b*	47%
5 a	100	*b*	50	*c*	$\frac{1}{10}$	*d*	$\frac{37}{100}$	*6a*	66	*b*	3·4	*c*	4	*d*	$\frac{7}{33}$

Arithmetic—Answers to Revision Exercises

Page 222   Revision Exercise on Chapter 1

1 a	64	*b*	144	*c*	0·49	*d*	1·96	*e*	0·0025
2	*a, d, e*			*3*	25, 36, 49, 64, 81				
4 a	9 cm²	*b*	16 mm²	*c*	3600 m²	*d*	33·6 m²	*e*	625 km²
5 a	5·8	*b*	44·9	*c*	68·9	*d*	3·1	*e*	30·8
6 a	5·29	*b*	47·61	*c*	76·21	*d*	34·34	*e*	82·63
7 a	807	*b*	207 900	*c*	0·1156	*d*	0·0605	*e*	0·007 569
8 a	3	*b*	12	*c*	40	*d*	15	*e*	2·5
9 a	6 m	*b*	50 cm	*e*	1·2 mm	*d*	17 m	*e*	1·4 cm
10	140 cm	*11*	*a, d*						
12a	9, 10	*b*	5, 6	*c*	2, 3	*d*	20, 21		
13a	3·6	*b*	1·7	*c*	7·6	*d*	8·4	*e*	5·6
14a	13·1	*b*	53·9	*c*	0·959	*d*	2·41	*e*	0·259
f	48·9	*g*	0·0469	*h*	1·04	*i*	32·9	*j*	104
15a	3·32	*b*	4·24	*c*	9·43	*d*	2·51	*e*	1·57
16a	5·1962	*b*	8·5440	*c*	2·4083	*d*	1·2728	*e*	9·7980
17	90 m	*18*	343 cm³	*19*	25, 64, 400; 1000th				
20a	26·5	*b*	83·7	*c*	837	*d*	0·265	*e*	0·837
21a	63·2	*b*	0·632	*c*	20	*d*	0·2	*e*	632;
	c and *d* are exact.								
22	14 cm, 56 cm; 28 cm			*23*	2·24 cm, 4·48 cm				
24a	53·29, 2·70	*b*	5329, 8·54	*c*	0·36, 0·775	*d*	360 000, 24·5		
25	1·5	*26a*	7	*b*	1·87	*c*	2·52		
27a	243	*b*	100 000	*c*	316				

Page 224   Revision Exercise on Chapter 2

1 a	8	*b*	3	*c*	200	*d*	0·1	*e*	0·2	*f*	2
g	200	*h*	4000	*i*	2000	*j*	0·5	*k*	0·3		
2	117	*3*	0·649	*4*	5·62	*5*	0·143				
6	250 000	*7*	1·37	*8a*	62	*b*	23	*c*	19	*d*	15
9	1·04	*10*	598	*11*	0·006 14	*12*	0·003 72				
13	0·459	*14*	1·66	*15*	1·20	*16*	0·347				

Answers

Page 225 Revision Exercise on Chapter 3

1	440 m, 15400 m²	2	157 cm, 94p	3	2.41×10^6 km
4	154 m², 8 kg	5	60 cm, 192 cm²	6	13·9 cm, 9·8 cm²
7	1260 cm²	8	78 m²	9	110 cm, 110 m
10	352 m/min, 264 m/min	11	85 m²	12	370 cm²
13a	22·8 cm b 7·08 m	14	56 m	15	14 cm, 616 cm²
16	10·5 mm	17	270, 21·5%	18	97 cm²
19	28300 km/h	20	9300 km		

Page 227 Revision Exercise on Chapter 4

1 b 12, 20, 10   2a 180   b 14%   3a 22   b 84%
4 a 2·8   b $\frac{11}{72}$ 5b 100, 83·15   c(1) 22% (2) 19%
6 a 3, 2   b 4·00, 4·55   c 59%, 59%   d 12%, 22%
7 Probability: $\frac{1}{36}$ $\frac{1}{18}$ $\frac{1}{12}$ $\frac{1}{9}$ $\frac{5}{36}$ $\frac{1}{6}$ $\frac{5}{36}$ $\frac{1}{9}$ $\frac{1}{12}$ $\frac{1}{18}$ $\frac{1}{36}$ Mean = 7
   Frequency: 4   8 12 16 20 24 20 16 12   8   4
8 a 1·2   b 1·9, 1·1   c 1·7   9 28   10 11 km/litre; 1·1p
11a 4·7 cm   b 2·9 cm

Arithmetic—Answers to Cumulative Revision Exercises

Page 239 Cumulative Revision Exercise A

1 a £61·20   b 62 p   2a $2 \times 3 \times 7$ b $2^3 \times 3$ c $2^3 \times 3 \times 5$
3 a 0, 8, 16, 24, ..., 96   b 0, 14, 28, ..., 98   c 56
4 a 19, 17   b 24, 18   c 70, 70   d 120, 120; no, yes
5 a $\frac{1}{3}, \frac{3}{8}, \frac{5}{12}$ b $\frac{7}{33}, \frac{7}{32}, \frac{7}{31}$ 6a 7   b 35   c 105   d $\frac{7}{32}$ e $3\frac{1}{16}$
7 a 0·0036   b 4·58
8 a 3.27×10^2 b 4.5×10^4 c 8×10^6 d 9.236×10^2
9 a 4   b 19·9 km   10 £7·67   11 4·7 kg   12 £4·62, £6·93
13 440, 40, £22·04   14 £98·40   15 1640 cm³
16 480   17 b and d   18 16·9 cm², 16·4 cm
19a 8 km/l   b 0·5 unit/h   c 108 litres/min   20 £1·47
21 ——   22 65·71 marks   23 700 m by 300 m; 21 ha
24 1·52 m   25a 42, 58, 21 p   b 9·6, 3·3, 7·9 pesos
26 11111, 100101   27 100, 1011, 11010
28a 10000, 100000; 1, 2, 4, 8, 16, 32   b 10101, 101010; 1, 2, 5, 10, 21, 42
29 100   30 £32·29   31 £131·28   32 £3·24
33 £272·10; 4p   34a £16·15   b £14·82   c £33·72   d £22·63
35 $3\frac{1}{3}$%   36 £10·82   37 633·20 fr, 287·65 fr   38 8.64×10^7 km
39 35 min   40a $\frac{1}{2}$ b $\frac{1}{2}$ c $\frac{2}{3}$ d $\frac{1}{2}$ e $\frac{2}{3}$
41a 93   b 3   42a $\frac{1}{3}$ b $\frac{3}{4}$ c 0   d 1
43a 88 cm, 616 cm²   b 18·8 m, 28·3 m²   44 898 cm²
45a 10 h 52 min, 11 h 34 min, 11 h 9 min   b 1 h 28 min   c 57 km/h
46 480 km; 0·75p   47a 1707   b 17 p   48 £22·40
49 4; 15   50 75

Page 244 Cumulative Revision Exercise B

1 $2^3 \times 3^2 \times 7, 2^2 \times 3^3 \times 5; 2^2 \times 3^2; 2^3 \times 3^3 \times 5 \times 7$
2 a yes; it is commutative   b associative   c yes; 6
3 a $\frac{7}{33} < \frac{7}{31} < \frac{8}{31}$ b $\frac{4}{15} < \frac{3}{10} < \frac{1}{3}$ c $\frac{53}{64} > \frac{49}{60} > \frac{25}{31}$
4 $\frac{1}{10}$ a $\frac{7}{18}$ b $\frac{7}{24}$ c $\frac{38}{9}$
5 a 3.00×10^8 b 2.91×10^7 c 7.93×10^1 d 5.00×10^3

Answers

304

6	37·81	*7*	128; 1·6 cm	*8*	595	*9*	11	*10*	48 mm²
11	35 cm	*12*	728	*13*	*a* and *c*	*14*	$5·87 \times 10^3$		
15	2·80; 2·8036; 0·028036			*16*	0·02 cm²	*17*	£17·92		
18	1 h 52 min; 168 km			*19*	8·4 cm	*20*	5	*21*	——
22	200 m, 4 ha; 12·96 ha			*23*	8·10 g	*24*	100+0, 11+1, 10+10		
25a	1001010 m	*b*	101 m						
26	100011 < 100110 < 101010 < 101100 < 110001					*27*	1011010		
28	$\frac{1}{10}, \frac{1}{10 \times 10}; \frac{1}{10} + \frac{1}{100}, \frac{1}{10} + \frac{1}{1000}, \frac{1}{2} + \frac{1}{4} = \frac{3}{4}, \frac{1}{2} + \frac{1}{8} = \frac{5}{8}$								
29	£56·88, £4·74		*30a*	60%	*b*	$37\frac{1}{2}$%	*31*	£48·05	
32a	$24\frac{2}{3}$%	*b*	7·51%	*c*	$1\frac{1}{2}$%	*33*	£1·63	*34*	second by £1·50
35	10%	*36a*	$\frac{1}{4}$	*b*	$\frac{3}{4}; \frac{1}{8}$				
37	9, $\frac{1}{4}$	*a*	$\frac{1}{8}$	*b*	1	*c*	0	*d*	$\frac{3}{8}$		*38*	2·350 million
39a	$\frac{1}{8}$	*b*	$\frac{1}{8}$	*40*	717, 810 km/h	*41*	48 km/h	*42*	75·4 kg
43	3·1; 489000			*44a*	$\frac{1}{8}$	*b*	$1·7 \times 10^9$	*45*	482	*46*	243 g
47a	389 m	*b*	8830 m²; 10 cm by 6 cm, 38·9 cm, 88·3 cm²						
48	19 min 25 s			*49*	32 g of A, 6 g of C; 72 g			*50*	29 crowns

Page 250 True-False Revision Exercise

1	F	*2*	T	*3*	F	*4*	F	*5*	T	*6*	F
7	T	*8*	F	*9*	F	*10*	T	*11*	F	*12*	T
13a	T	*b*	T	*c*	T	*14a*	F	*b*	F	*c*	T	*d*	T
15a	T	*b*	F	*c*	T						

Computer Studies—Answers to Chapter 1

(Alternative answers are possible in this Exercise)

Page 256 Exercise 1

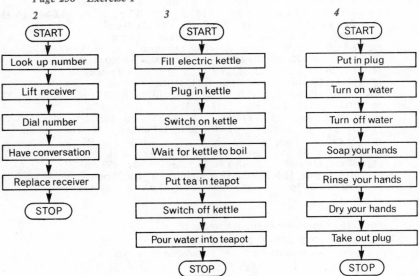

Answers

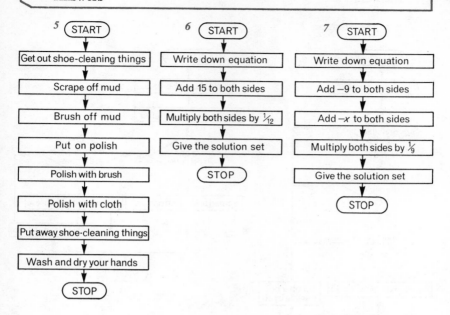

Page 257   Exercise 1B

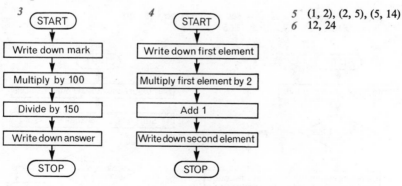

5 (1, 2), (2, 5), (5, 14)
6 12, 24

Answers

Page 259 Exercise 2

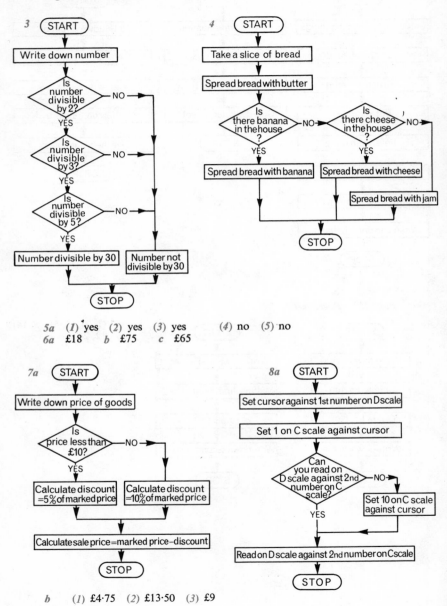

5a (1) yes (2) yes (3) yes (4) no (5) no
6a £18 b £75 c £65

b (1) £4·75 (2) £13·50 (3) £9

Answers

Page 261   Exercise 3

6

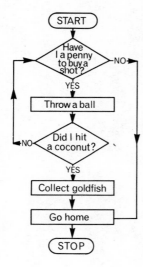